Ask the Astronaut

Ask the Astronaut

A galaxy of astonishing answers to your questions on spaceflight

Tom Jones

Smithsonian Books
Washington, DC

This book may be purchased for educational, business, or sales promotional use. For information, please write: Special Markets Department, Smithsonian Books, P. O. Box 37012, MRC 513, Washington, DC 20013

Published by Smithsonian Books
Director: Carolyn Gleason
Production Editor: Christina Wiginton
Editorial Assistant: Jaime Schwender

Edited by Jean Crawford
Designed by Service Station

Library of Congress Cataloging-in-Publication Data

Names: Jones, Tom, 1955 January 22- author.
Title: Ask the astronaut : a galaxy of astonishing answers to your questions
on spaceflight / Tom Jones.
Description: Washington DC : Smithsonian Books, [2016] | Includes
bibliographical references.
Identifiers: LCCN 2015037850 | ISBN 9781588345370
Subjects: LCSH: Space flight—Juvenile literature. Space flight—Physiological effect—Juvenile literature. Astronautics—Juvenile literature. Astronauts—Juvenile literature. Children's questions and answers. Outer space—Exploration—Juvenile literature.

Classification: LCC TL793 .J65 2016

DDC 629.4/1—dc23 LC record available at http://lccn.loc.gov/2015037850

Manufactured in the United States
20 19 18 17 16 5 4 3 2 1

It is difficult to say what is impossible, for the dream of yesterday is the hope of today and the reality of tomorrow.

—**Robert H. Goddard**

This book is dedicated to the next generation of space explorers, undaunted by the impossible.

Contents

Introduction

Space exploration has been at the core of my personal interests and professional life for more than five decades. When I was five, my grandmother gave me a slim book about spaceflight that seized my imagination and set me on a course toward the stars. As a student, I watched humankind's first steps on the Moon and made up my mind to be part of our push into the cosmos.

I was piloting Air Force B-52s when NASA unveiled its new space shuttle. A few years later I got my first glimpse of an orbiter—the *Columbia*—at its refueling stop in Tucson, Arizona. I worked hard to earn an opportunity to fly her, and ten years later I launched aboard that very ship on my third shuttle mission. I've been very lucky—and privileged—to represent the United States in space on four space shuttle missions, including a journey to the International Space Station.

Perhaps the most special moment of those adventures came near the end of my third spacewalk at the space station. I was ahead of schedule for my planned work outside, and was aware that I would soon have to return to the cabin of shuttle *Atlantis*. Taking a brief break from my work, I gripped a handrail at the very prow of the station and looked out at my surroundings.

The view through the plastic faceplate of my helmet was breathtakingly beautiful. I was perched on the bow of a giant, space-age vessel that was falling effortlessly around the globe. Above me, the station's golden solar array wings, against a velvet-black sky, stretched across the heavens like the canvas of a windjammer under full sail. A thousand miles ahead, the Earth's curved horizon ended in the thin, gauzy blue of the atmosphere. Two hundred and twenty miles beneath my boots, the brilliant blue waters of the Pacific, swept with dazzling white clouds, rolled serenely by. Struck by this vista, I felt tears of gratitude and humility welling in my eyes—what a gift to see this cosmic stage set by God, and how insignificant it made me feel!

Contemplating that scene was just one of many, many inspirational moments I experienced in space. Since my final mission ended, I've been trying to explain what living in space is truly like. Over the years I've spoken to hundreds of audiences and tens of thousands of people, from kindergarteners to corporate executives, from Air Force Academy classmates to church groups, and from international travelers to professional conferees. Every group has peppered me with questions, from the basic to the unusual. In this book I've answered more than 400 of my favorites to introduce you to the human experience in space.

My space story is interesting, but our future in space belongs to a new generation of twenty-first-century explorers. Thus, many of my answers are addressed to those preparing to explore, along with their parents, families, and teachers. Our drive into the solar system in the decades ahead will follow varied paths to new worlds, many glimpsed only by robot scouts. Where we go, when we go, and what innovations and discoveries we'll make in settling humanity in space are choices for these new pioneers. They will solve mysteries that have puzzled us for centuries, while asking—and answering—questions we have yet to imagine.

CHAPTER I

Aspiring to Space

An astronaut samples the surface of a near-Earth asteroid during a future deep-space mission. (NASA)

1. Did you always want to be an astronaut, and when did you decide?

From the first moment I learned about space exploration, I wanted to be part of it. As a ten-year-old Cub Scout, I visited the Martin Marietta rocket factory near my hometown of Baltimore, Maryland. The powerful Gemini-Titan II rockets being built there would take astronauts into space to rehearse the skills needed for the first Moon landing.

Here was the Space Race taking shape right in my home town! I remember looking up at those 10-story-tall, silver and black rockets and thinking that astronauts had the greatest job around. They were flying the most complex machines ever launched and going to places no humans had ever been. From then on, I began reading every scrap of information I could find about space exploration. I was determined to become an astronaut someday.

2. What encouraged you to become an astronaut?

When I was five, my grandmother gave me a copy of *Space Flight: The Coming Exploration of the Universe*. That book launched my curiosity about astronomy and rocketry. But what really got me hooked was the race to the Moon between the U.S. and the Soviet Union in the 1960s. As a ten-year-old already interested in airplanes and flight, I avidly followed each launch.

My teachers brought televisions into our classrooms so we could watch each Gemini and Apollo blastoff and splashdown. Our regular lessons were put on hold as we watched hours of space news. Television news anchors broadcasting the key moments of each mission were as excited as I was.

I recall the first American spacewalk, the first orbital docking between two spacecraft, and the first Apollo flight to circle the Moon. As the United States gained more and more experience in space, I saw how important the nation's progress in space was to my teachers, my parents, and my friends' parents, many of whom worked at the nearby Titan rocket factory.

I believed my parents and teachers when they told me: If you work hard in school, you, too, could be an astronaut one day. Watching the

A Titan II rocket launches Gemini 11 in 1966; the Titan II was assembled and tested at the Martin Marietta factory near Baltimore, Maryland. (NASA)

The gift from my grandmother that started my space career—1960. (Author)

1968 film *2001: A Space Odyssey*, quickly followed by the 1969 *Apollo 11* Moon landing, cemented my desire to work in space one day.

3. How did your parents react to your desire to become an astronaut?

My parents encouraged my early interest in astronomy and astronautics. They knew these would be good fields to get into because of the nation's growing interest in space exploration due to the Space Race with the Soviets. Also, many of our neighbors worked for Martin Marietta on the Gemini-Titan II program. The winter I turned 12, my folks bought the family a 3-inch reflector telescope for Christmas. I put it to constant use examining the face of the Moon and the nearby planets.

When I was in high school, I was eager to start on the path to becoming a test pilot and getting an education in astronautics. My father encouraged me to meet with representatives of the Air Force Academy to pursue my interest in studying there after graduation. My parents didn't laugh at my dreams. Instead, they told me that with hard work I could succeed. My mom, who was afraid to fly and chose never to take wing in an airplane, kept her fears to herself as I pursued one of the riskiest lines of work imaginable.

4. What was your career path to becoming an astronaut?

When I first decided to become an astronaut, a few of NASA's space fliers were scientists, but most were test pilots and engineers. So I joined the Air Force to give myself the best chance of earning my test pilot credentials. After graduating from the Air Force Academy, I qualified as a pilot and flew for five years as a B-52 Stratofortress bomber copilot and aircraft commander.

When NASA introduced the space shuttle, I learned that I could qualify as a mission specialist astronaut by earning an advanced science degree. Because my favorite subjects were astronomy and space science, I decided to pursue a degree in planetary science. At the University of Arizona, I spent five years earning my doctorate, specializing in asteroid research.

My first job after getting my PhD was with the Central Intelligence Agency (CIA), working as a program management engineer. That experience showed I could work successfully in a demanding research and development position, similar to the challenges I'd face at NASA. My last job before joining the astronaut corps was as a senior scientist with Science Applications International Corporation, helping NASA plan its solar system exploration program.

5. How did you succeed in reaching your goal of becoming an astronaut?

Knowing that I would face tough competition, I worked hard to get top grades in college and excel in all of my jobs. I wanted to be an outstanding bomber commander and the best pilot in the Air Force. After that, I wanted to be a well-regarded scientist. Although I originally wanted to become a test pilot, I decided that space science was more interesting because the universe always offers new discoveries. My goal was to become an astronaut, but if I didn't reach that goal I wanted a career that would keep engaging and challenging me over the long run.

Each time I applied to be a NASA astronaut, I tried to add new talents and skills to my resume to show steady improvement. I was turned down twice, but I persisted. One astronaut colleague applied thirteen times before being selected. Determination counts!

6. How old were you when you went into space?

I was 39 during my first mission and 46 the last time I flew. According to Albert Einstein's general theory of relativity, time slows down for people moving rapidly through space, so they age more slowly than people on Earth. But all my time in orbit traveling at 17,500 miles per hour (28,530 kilometers per hour) shaved less than three milliseconds off my age!

7. How many times have you been in space?

I was privileged to represent the United States off the planet on four space shuttle missions. The first three were scientific expeditions. The last mission contributed to the assembly of the International Space Station by delivering and activating the $1.4 billion U.S. science laboratory module, *Destiny*.

8. What was the purpose of each of your four missions?

Each shuttle mission gets a number preceded by "STS," which stands for Space Transportation System. That's what they called the shuttle when it was being designed, and the acronym just stuck. The first shuttle mission was STS-1. The four missions I flew on were:

- STS-59, which examined our changing planet with the *Space Radar Lab 1* Earth imaging system.
- STS-68, which used *Space Radar Lab 2* to scan Earth for natural and man-made change.
- STS-80, which launched and retrieved two scientific research satellites.
- STS-98, which delivered the U.S. *Destiny* science lab to the International Space Station.

9. How many times have you orbited the Earth?

With my crews I completed 847 revolutions, or orbits, around Earth. We traveled 21.7 million miles, or almost a quarter of the distance from here to the Sun. Surprisingly, I never got farther from Earth than about 233 miles (377 kilometers). The Apollo astronauts on their journeys to the Moon traveled a thousand times farther from our planet than I did.

10. How much time have you spent in space?

Altogether, my four missions totaled 53 days, 49 minutes off the planet. The record for the longest single space mission belongs to Valeri Polyakov, who flew on Russia's *Mir* space station for 438 days, from January 9, 1994 to March 22, 1995. Russia's Gennady Padalka, with his five stays aboard the *Mir* and International Space stations, has lived in space for 879 days so far, longer than any other human.

CHAPTER II

Training for Space

A Delta IV rocket lifts the Orion spacecraft on its first unpiloted test flight in December 2014. (NASA)

1. How do you simulate launch accelerations and weightlessness for astronauts in training?

I couldn't wait to experience the sensations and sights of spaceflight—who wouldn't? But first I had to face them in training.

To get acquainted with the strong acceleration forces, also known as g-forces, that the human body must endure during launch, I flew with shuttle astronauts in our Northrop T-38 trainer jets through wrenching aerobatics and high-g maneuvers—the kind of flying I loved as an Air Force pilot. I also trained in a large human centrifuge, a machine that spins you around very fast to create these acceleration forces.

My most exciting launch simulation was a trip to the centrifuge at Brooks Air Force Base in San Antonio, Texas. This centrifuge whirls a cabin around at the end of a long steel arm, subjecting its occupant to the same kind of acceleration you feel when an airliner powers up its engines during takeoff, squeezing you back into your seat. Suited up for liftoff, I was strapped to a shuttle seat and subjected to three simulated launches each lasting 8 minutes, 30 seconds. For the last minute, the force on my body was so strong that I felt as if I weighed nearly 500 pounds, three times my normal weight. It was hard to breathe or raise my arms, but I now knew what to expect during the ride to orbit.

To experience the weightless sensation of free fall, I boarded NASA's KC-135 jet, once known informally, but accurately, as the "Vomit Comet." Each exhilarating flight included 40 roller-coaster-like climbs and descents, creating free fall conditions for up to 25 seconds at a time. The repeated transitions between free fall and twice the normal force of gravity can make you feel sick to your stomach, sometimes very sick. This plane always delivered on its nickname!

2. Does NASA operate an anti-gravity room in Houston, a room in which gravity can be turned off?

Tour guides at NASA's Johnson Space Center are often asked this question, but there is no such room. Instead, NASA relies on its specialized high-speed jets to simulate the weightless feeling of free fall, and uses the Neutral Buoyancy Lab—a mammoth swimming pool near the

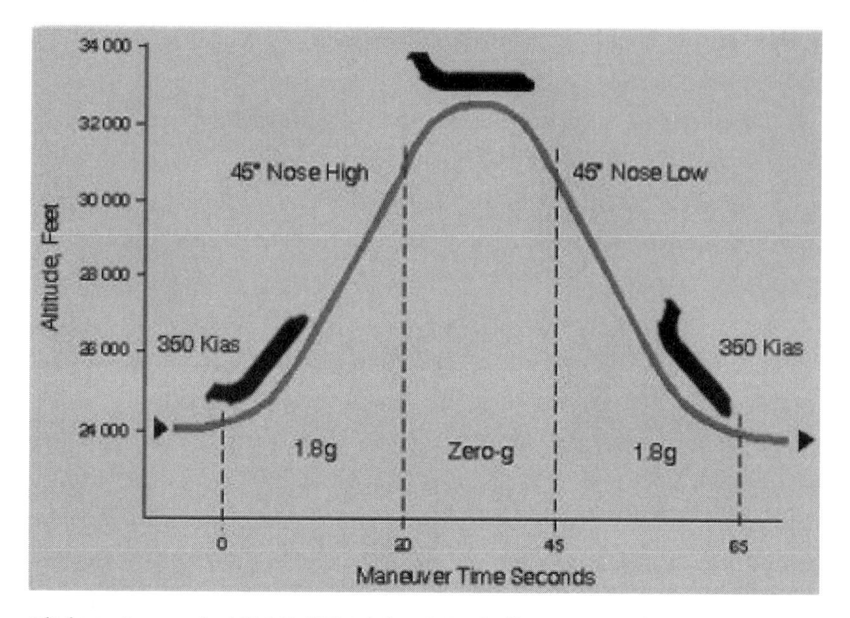

Flight trajectory for NASA's "Weightless Wonder" C-9 jet, used to expose experiments and astronauts to "zero-g" or free fall conditions. (NASA)

Space Center—to train astronauts on how their tools and spacesuits will behave while walking in space.

3. How do astronauts learn to operate a spaceship?

Flying a spaceship seems enormously complicated at first, but NASA broke down my training into manageable steps that occurred over a period of about one year. We started with introductory classes aimed at understanding the shuttle's systems (electrical, hydraulics, computers, propulsion, etc.).

Next, we practiced operating those systems with checklists and simple training replicas of our switch panels and displays. Then we moved on to "flying" the space shuttle in a simulator with all systems working together. Finally, once assigned to a mission, my crew practiced that mission in the simulator until we knew how to perform every aspect of it and respond crisply to any emergency. That whole process took about two-and-a-half years. Astronauts assigned to the International

European Space Agency astronaut Samantha Cristoforetti trains on an International Space Station mockup in the six-million-gallon pool of NASA's Neutral Buoyancy Lab near Johnson Space Center in Houston, Texas. (NASA-ESA)

NASA astronaut Scott Kelly trains inside a Soyuz simulator at the Gagarin Cosmonaut Training Center (GCTC) in Star City, Russia. (NASA/Bill Ingalls)

Space Station train in a similar step-by-step approach to operate their orbiting lab and space transports.

4. What languages do astronauts need to know to work on the International Space Station?

Just as English is the language of international air transportation, it is also the official operations language of the ISS. Crewmembers must be able to use English for handling all tasks related to operating the station. Because half the crew is from Russia, with frequent communications to Russia's Mission Control in Moscow, ISS astronauts must also learn to speak Russian well enough to conduct technical tasks in that language. Learning Russian is also necessary for operating the *Soyuz* transport spacecraft. Besides, knowing your crewmates' language is the best way to build trust and friendship during years of training and months in orbit.

5. What jobs do astronauts do, and what are their job titles?

On the International Space Station, astronauts conduct scientific research and operate the orbiting outpost for tours of duty lasting about six months. They perform spacewalks to maintain and repair the space station, and operate the robot arm (Remote Manipulator System) to grapple arriving supply vehicles and install new scientific equipment. On the ISS, all astronauts are designated as flight engineers, but only one serves as commander. This is also the case on space taxis headed to the ISS, like the *Soyuz* or the *Crew Dragon* and *CST-100 Starliner* commercial craft. NASA gives all of its qualified spacecraft crewmembers the job title astronaut, even before their first flight.

6. How do astronauts learn to handle emergency situations?

Learning how to respond correctly to a life-or-death emergency is the toughest challenge astronauts face in spaceflight training. Our instructors deliberately challenged our knowledge and skills. Once we were proficient at flying the shuttle simulator, they would introduce one failure after another until we made a mistake. They'd do this not just to make us sweat, but to teach us how to fly the orbiter safely while trying to recover from an emergency, like a failed engine or cabin fire. We learned to share the workload to keep such problems from overwhelming our team. These fraught-with-failure simulator sessions truly tested our ability to keep our cool and work through emergencies as a crew.

7. How many astronauts are in a space crew?

A crew on the International Space Station is made up of six astronauts, but can expand to a crew of seven. Russia's *Soyuz* capsule normally launches with three crewmembers. China has flown three astronauts in its *Shenzhou* capsule. The shuttle flew with as few as two and as many as eight crewmembers. In the future, commercial space taxis may carry as many as seven crewmembers and passengers, and the *Orion* spacecraft will carry as many as four astronauts to deep space on missions lasting up to a month.

Crewmembers pose for a group photo aboard the ISS in 2007. Pam Melroy, commander of shuttle mission STS-120, is at the center. At bottom in dark shirts, from left, are ISS expedition 16 crewmembers Clay Anderson, commander Peggy Whitson and Yuri Malenchenko. Clockwise from left, STS-120 astronauts Stephanie Wilson, Dan Tani, Scott Parazynski, Doug Wheelock, Paolo Nespoli and George Zamka. (NASA)

8. Who designs the crew's mission patches?

The crew does. Once assigned to a flight, the crew meets to discuss the themes of the upcoming mission and how to represent them in an attractive design. The astronauts might sketch several ideas for patch designs, and many ask for suggestions from artist friends or colleagues.

The author's STS-98 space shuttle crew developed the original outline design and color scheme for their mission patch, showing Atlantis heading for the ISS with the new United States Destiny science lab. (NASA)

Once they decide on an attractive design, NASA and its partner agencies produce a draft emblem and circulate it for approval. The completed patch then represents the hopes and goals of the crew on this new adventure.

9. How many years will an astronaut have to wait before his or her first flight?

For NASA's astronauts selected to fly to the International Space Station, a wait of five years or more from the time they are hired is not unusual. Each spends one to two years training as an astronaut candidate, a year or more in a technical assignment as a member of the astronaut corps, and then two to three years of expedition-specific training once assigned to a flight.

After an ISS expedition, astronauts typically wait another five years before heading to space again. They take on an Earth-bound job to give them time to readapt and recharge before starting to train again. Because NASA will fly only about six astronauts per year through the mid-2020s—on the ISS, in space taxi tests, and on the *Orion* spacecraft—this initial wait is not likely to get any shorter.

10. How long does an astronaut train for a specific mission or expedition?

Astronauts train for at least two years to get ready for an expedition to the International Space Station. Those flying highly automated commercial space taxis will need only a year or so of training for their brief trip to and from the ISS. According to NASA, training for long duration missions on the ISS is very arduous and takes from two to three years. This training requires extensive travel, including long periods in other countries training with our international partners. Training for future deep-space expeditions, involving multiple ships like the *Orion* spacecraft, and possible surface exploration of an asteroid or planet, might take up to five years.

Artist concept of NASA's Orion deep-space crew vehicle in Earth orbit. (NASA)

11. What does astronaut candidate training consist of and how long does it last?

Astronaut candidate training prepares you to do your job effectively in space, to cope with the challenges of the space environment, and to respond effectively to emergencies that may occur. You'll begin with classroom instruction, and then move on to high-performance jet flying, survival training, exposure to free fall, intensive spacecraft simulator sessions, food preparation classes, Russian language classes, physical conditioning, maintenance training, spacewalk preparation, and leadership training. I found astronaut candidate training to be fast-paced, intense, challenging, and fun. I was sustained and motivated throughout by the idea that at the end of it all, I would experience spaceflight!

12. Why is aviation training good preparation for spaceflight?

Spaceflight Readiness Training is what NASA calls its use of jet trainers to prepare astronauts for space missions. There are a lot of similarities between the decisions pilots make in a jet cockpit and those an astronaut must make in a spacecraft. In the air, astronauts must continually make decisions about safety, mission priorities, weather, fuel levels, and flight routes. They must respond to possible failures in aircraft systems or other emergencies. They also must communicate clearly with their crewmates and with air traffic controllers via radio.

I flew the Northrop T-38N jet trainer, paired with NASA instructor or astronaut pilots. I also commanded NASA's Cessna Citation II business jet to give new astronauts practice in decision-making and teamwork during flight. Making the right choices and communicating clearly while safely flying a high-performance aircraft in a stressful, sometimes hostile environment—where the wrong decision can mean injury or death—is very similar to the intense, dynamic nature of spaceflight.

Astronauts Eileen Collins and Tom Jones with their Northrop T-38 jet trainer, on a visit to the Palmdale, California factory where the space shuttle orbiters were built. (NASA)

13. How do you prepare for spaceflight's physical demands?

Spaceflight is a physically demanding job, and being fit helps astronauts withstand the rigors of leaving Earth, working in space, and returning home.

To keep in shape, astronauts work out regularly in a gym. For many years, the astronauts used a modest but effective gym at the Johnson Space Center. It had a half-basketball court, a couple of squash and racquetball courts, fitness machines, a climbing wall, and an outdoor running track. Astronauts now use the large Columbia fitness and rehabilitation center nearby (named after the shuttle crew lost in 2003), with modern exercise equipment and a rehab swimming pool. There are also offices for the fitness coaches who prepare crewmembers for long-duration spaceflight and advise them during rehabilitation.

One year before launch, crewmembers are assigned a strength and conditioning specialist who designs a training program that emphasizes all-around fitness and regular strength training. The astronauts follow this program during workouts several times a week.

Before launch, each astronaut's heart and lung fitness is tested on a cycle ergometer, a stationary bicycle with sensors that can measure cardiovascular performance. They are also tested for functional fitness, agility, and isokinetic strength. When the astronauts return from their mission, their physical fitness is tested again and compared to what it was before launch.

14. What facilities are used to train astronauts in Houston?

The Johnson Space Center in Houston, Texas, where NASA astronauts work and train, houses classrooms, computer-based trainers, mission simulators, and mock-ups of the International Space Station (ISS), *Orion* spacecraft, and commercial spacecraft. Simulators let astronauts practice tasks they will have to do in space, such as docking a vehicle at the ISS or flying or driving over the surfaces of asteroids and the Moon. At Mission Control, astronauts are exposed to real space operations by working alongside experienced flight controllers as they communicate with astronauts in space.

Outdoors, a mock Mars landscape gives astronauts a taste of driving a rover and exploring the geology of that planet. Just north of the space center is Ellington Field, where astronauts receive spaceflight readiness training in a fleet of T-38 Talon jet trainers. Also at the airport is the Neutral Buoyancy Lab, a six-million-gallon pool used for spacewalk training.

15. How realistic are the simulators used by NASA to train its crews?

There are two simulators for the International Space Station at the Johnson Space Center. They don't move and they can't simulate free fall,

The Space Vehicle Mockup Facility (SVMF) is a full-scale, high-fidelity replica of the International Space Station layout. (NASA)

but they do a good job of reproducing the look and feel of the station, and they give crews the skills they need to operate their orbiting outpost.

In full-sized mockups of the station's modules, astronauts get realistic training in maintaining the station, fighting fires, and transferring cargo between the station and visiting spacecraft. They also gain experience operating safety equipment, hatches, and spacesuits.

Another mission simulator allows crews to practice using the ISS computer systems to communicate, operate systems on board, maneuver the station, and practice docking procedures.

Soon-to-be-added simulators for commercial space taxis will put crews in the liftoff position, lying on their backs to practice countdown, ascent, and splashdown procedures.

16. Do other countries train astronauts and where?

Each International Space Station partner has its own astronaut training facilities. Japan's is located at Tsukuba Science City, outside Tokyo; Europe has the European Astronaut Centre near Cologne, Germany; Canada's is the John H. Chapman Space Centre in St. Hubert, Quebec; and Russia has the Gagarin Cosmonaut Training Center in Star City, near Moscow.

After initial training, Japanese, European, and Canadian astronauts move to Houston for their specific ISS expedition training. NASA astronauts spend nearly half of their time in Moscow during ISS expedition training, and Russian cosmonauts spend the same amount of time training in Houston. China trains its astronauts at the Astronaut Center of China in Beijing, operated by the Chinese air force.

17. What other space training programs are available to aspiring astronauts?

There are a number of places around the U.S. and in Russia that offer a taste of spaceflight training. For example, the Kennedy Space Center Visitor Complex in Florida has the Astronaut Training Experience, while the U.S. Space and Rocket Center in Alabama runs a United States Space Camp for young people and adults. In Russia, the company Space

Adventures, Ltd. runs a program that lets people experience spaceflight training at the cosmonaut training center near Moscow.

18. Is Space Camp a good idea for future astronauts?

The space camps I've visited offer a brief but realistic taste of an astronaut's training and flight experiences. Many of the activities give campers hands-on experience in simulators. This allows them to experience some of the things astronauts do in space, such as manipulate a robot arm, fly a spacecraft, or work on experiments at the International Space Station. Campers learn how to work as a team under demanding conditions and get a taste of how much fun spaceflight can be.

To get a closer look at an astronaut's duties and a chance to meet others with similar interests, consider space camp as a fun, educational vacation option.

19. Are private space firms training their own astronauts?

Commercial firms have developed their own training programs, or have hired specialists in spaceflight training to do the job for them. The National Aerospace Training and Research Center near Philadelphia offers professional training for commercial spaceflight crews and passengers. This training includes experiences in simulators, a centrifuge, and an altitude chamber. Virgin Galactic promises its passengers spaceflight training at Spaceport America in New Mexico.

20. Were you well prepared for your first venture into space?

As daunting as spaceflight seems to many of us Earth-bound humans, our half century of experience in space has given us the ability to train astronauts well for its challenges. NASA did an excellent job of preparing me for my first shuttle mission.

When *Endeavour* reached orbit and its engines shut down, I felt completely comfortable in my spaceship. I took off my spacesuit glove and watched it tumble gently in front of my face. As I unstrapped myself from my seat on that first trip, I looked around and realized how familiar everything was. I knew every switch, storage locker, and system, and I knew exactly what I had to do. Within minutes I was assisting my crew in getting our successful *Space Radar Lab* operations underway. It was amazing to be doing what I'd rehearsed so many times—but now I was actually working in space!

21. Is astronaut training fun?

I can truthfully say that most of my astronaut training was out-and-out fun. Sometimes it was tiring, as underwater spacewalking training always was. It could be stressful, as were most simulator sessions. And it could be difficult, as was mastering the shuttle's computer and propulsion systems.

But I also had a lot of really fun experiences, such as flying a jet, wriggling into a spacesuit, trying out space food, practicing putting together a space station, riding the "Vomit Comet," being spun in a centrifuge, and controlling a robot arm to retrieve a satellite from orbit. One of the best parts was working with bright, motivated people who would fly with me, help me in space, and become my friends for life. Who *wouldn't* want to go to work each morning?

There is one drawback to all this fun and hard work. There's so *much* training, particularly as launch day approaches, that travel and long hours often keep you from your family. While often fun, training was surely the most challenging part of my job as an astronaut.

22. What were your best and worst moments in astronaut training?

My best astronaut training moment was the first time I donned a spacesuit and was lowered into NASA's Weightless Environment Training Facility. This was the 25-foot-deep swimming pool NASA used until 1996 to teach astronauts how to maneuver their spacesuits and use their tools in the free fall, or weightless, environment of orbit.

I remember looking up at the glittering surface of the water and seeing the divers' bubbles gently floating upward. The only sound I heard was the whisper of air flowing through my helmet and water gurgling outside my suit. I knew I wasn't really in space, but also knew I was truly training for a spacewalk.

My worst moments occurred in that same pool, because training for a spacewalk can be exhausting. After six hours underwater I was often mentally and physically spent. Going topside and getting out of the spacesuit was a blessed relief. Astronaut training was always mentally challenging, frequently demanding physically, but nearly always fun. And I was taught by wonderfully talented instructors and astronauts.

The author about to head underwater for space station spacewalk training at Tsukuba Space Center in Japan. (NASA)

23. Why is the NASA human spaceflight center in Houston?

In 1961, NASA announced a nationwide search for its new Manned Spacecraft Center. The new site had to meet the following criteria: be near water transport and an all-weather airport, be closely linked via telecommunications to the rest of the nation, have adequate water for test facilities, possess a mild climate for work outdoors, have access to a capable workforce, and be located near a culturally attractive community.

In September 1961, NASA announced that the new center—renamed the Johnson Space Center in 1973—would be built in Houston. To house the center, Rice University donated 1,000 acres of undeveloped land located 25 miles (40 kilometers) southeast of Houston, near Galveston Bay. Vice President Lyndon B. Johnson, a former U.S. Senator from Texas, U.S Representative Olin Teague of Texas, and others in the state congressional delegation undoubtedly had something to do with the selection. Eight years later, "Houston" was the first word transmitted to Earth by astronaut Neil Armstrong from the surface of the Moon.

24. How do astronaut candidate classes choose their nicknames?

Once a new astronaut candidate class arrives at the Johnson Space Center for training, the group is traditionally given a class nickname by colleagues in the astronaut corps. Group 14 was originally called the Pigs, but finessed a name change to the Hogs. Group 16 was the largest class at the time, with 44 members, and so crowded NASA's training facilities that it was called the Sardines. The 2009 class campaigned to be named the Chimps, but fellow astronauts designated them "the Chumps" instead, perhaps because its members faced a long wait before they would get their chance to fly to the International Space Station.

25. How do you train for survival on Earth in case of an emergency landing?

In the case of a launch abort or an emergency descent from orbit, astronauts must be able to survive on the ground for 24 to 48 hours until rescuers can reach them. Crews must be prepared to use their survival gear and act quickly as a team no matter what environment they land in—desert, icy terrain, or the open ocean.

Crews practice getting out of the descent module on land and in water, and getting out of spacesuits and into anti-immersion, open-ocean survival suits or cold weather gear. They learn to deploy a life raft and practice being hoisted into a rescue helicopter. They also are trained in basic survival skills, such as building fires, signaling, constructing shelters, performing first aid, and finding and cooking food in the wilderness.

26. How did you rehearse the launch of your space shuttle?

My crews "flew" two kinds of space shuttle mission simulators to practice for our actual launches into orbit.

European Space Agency astronaut Alexander Gerst exits a Soyuz spacecraft mockup during winter survival training near Star City, Russia, in January 2013. (Gagarin Cosmonaut Training Center)

The "motion base" simulator used hydraulic jacks to shake and tilt the flight deck just as it would move during launch. The vibration and accelerations weren't as strong as the real thing, but it taught us what normal changes in acceleration or vibration would feel like. The inside of the simulator was an exact replica of the space shuttle's cabin. All the switches and displays worked just as they would during launch, and allowed us to practice handling all kinds of serious launch emergencies.

The "fixed base" simulator didn't move, but it included a middeck compartment so the entire crew could rehearse our tasks together, from ascent into orbit and through our mission.

Three weeks before launch, we flew to the Kennedy Space Center in Florida for a launch dress rehearsal. We practiced the launch countdown while strapped in our seats in the real shuttle crew cabin, doing everything except actually starting the engines.

27. What changes in training will be needed for future journeys into deep space?

Many of the current training techniques will work equally well when preparing astronauts for journeys to the Moon, asteroids, and Mars. But we will use more computer simulations in addition to the

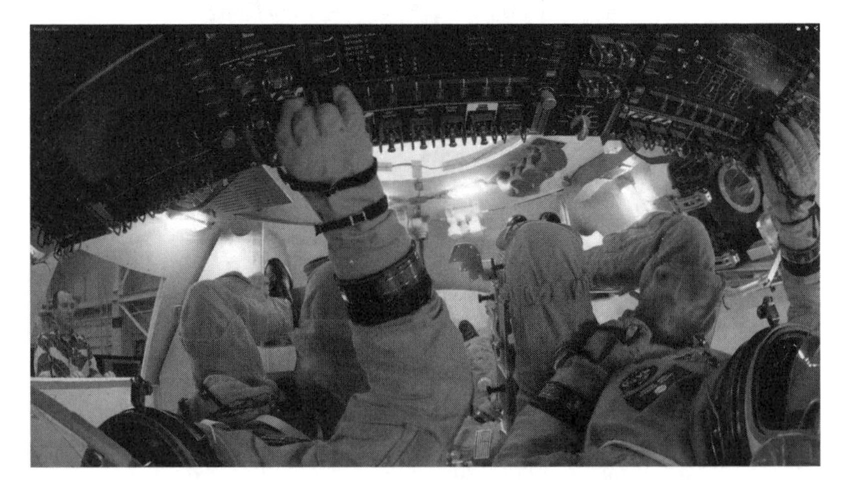

NASA astronauts evaluate the computer and control layout in a mockup of the Orion spacecraft. (NASA)

realistic physical simulators. This is because controlling ships designed for deep-space travel will be done with as few as two or three modern computer displays rather than hundreds of physical switches.

Crews that travel far from Earth will have to be much more self-sufficient than astronauts in low-Earth orbit. In the case of emergencies, they won't be able to depend on instant help from Mission Control, because the farther you are from Earth the longer it takes for a message to get to you. If you are on Mars, for example, it might take as long as 40 minutes to receive an answer to a question. Crews on these lengthy voyages will also receive refresher training using videos and lessons radioed from Earth. Astronauts on the International Space Station are already testing many of these training techniques to make sure future explorers are ready.

CHAPTER III

Getting to Space

Shuttle Columbia lifts off with the author and crew for a record-breaking 18 days in space on mission STS-80. (NASA)

1. What do astronauts feel at liftoff?

The Soyuz rocket that takes crews to the International Space Station is a three-stage launcher based on a Soviet intercontinental ballistic missile. After engine ignition, the Soyuz takes over 20 seconds to build up enough rocket thrust to lift off, subjecting the crew on board to a growing, rumbling vibration.

Once clear of the launch support structure, the rocket accelerates smoothly for two minutes, accelerating the astronauts at twice their normal weight—2 g's of acceleration. At 114 seconds after liftoff, the escape rocket at the top is no longer needed and fires a small rocket motor to clear it away from the main rocket. Three seconds later, the four first-stage boosters drop away with a loud bang. The second stage engines continue accelerating the crew faster and higher into space, pushing the crew back into their seats with four times the normal force of gravity—4 g's.

At 2 minutes, 37 seconds after liftoff, a startlingly loud bang and bright flash outside announces the separation of the shroud that protected the spacecraft in the atmosphere. Four minutes, 47 seconds after liftoff, the second stage engines cut off and the empty rocket stage drops away. At separation, the crew is thrown forward into their straps for a moment of free fall, and then shoved back hard into their seats as the third stage ignites. Astronauts liken the sensations of third stage ignition to riding in the middle of a flying car wreck.

Finally, at 8 minutes, 44 seconds after liftoff, the third stage shuts down and vaults the crew into orbital free fall. Four seconds later, the *Soyuz* separates from the third stage and the crew is on its way to the ISS.

2. Why does NASA put crews in medical quarantine before launch?

Astronauts undergo a week of medical quarantine before launch to limit their exposure to cold viruses and other common diseases they might catch from their families, co-workers, or the public. Only essential co-workers who have been screened by the flight surgeons are permitted to visit the crew. Because of all the germs floating around schools, astronauts cannot kiss or hug their children or come within 10 feet of them during the week before launch. Spouses are able to visit astronauts in quarantine, but only after they pass their own physical exam. The quarantine strategy works. Very few launches have been delayed due to sickness, and astronauts rarely get sick once they reach orbit because the spacecraft contains few infectious viruses or bacteria.

3. Before launch, how do you prepare your body for your work schedule in space?

Our sleep-wake schedule in space is set by major mission events, such as liftoff, launching and retrieving satellites, docking with the International Space Station (ISS), and landing. Flight surgeons want the crew to be refreshed and wide awake during these events, so we wake up about five or six hours before they are scheduled to take place.

Astronauts use the quarantine period before launch to adjust their wake and sleep times to match the new orbital schedule. To adapt more quickly, I would sleep in a pitch-dark room to reset my brain to the new nighttime, and I'd work under brilliant lights in crew quarters to adjust to the mission's day. We even wore dark welder's goggles when we had to go outside in daylight, to drive to the simulator or to our offices, so our brains still thought it was nighttime. It took me about four days in quarantine before my body felt comfortable with the new orbital work schedule.

This light therapy technique was especially effective for my first two missions when I had to shift my normal Houston bedtime about 12 hours to adapt to the night shift for *Space Radar Lab* operations.

4. What is a launch window and why is it so short when launching a spacecraft to the International Space Station?

To rendezvous with an orbiting body like the ISS or the Moon, the spacecraft's launch must be carefully timed so that its orbit matches that of the target, allowing the incoming spacecraft to dock or land.

For a spacecraft to dock with the ISS, launch controllers must wait until the Earth's rotation brings the launch pad under the orbit of the station. This alignment lasts for just a few seconds or minutes, depending on the ship's ability to steer and compensate for any launch delay. This alignment interval is called the launch window. If the weather is bad, or a spacecraft or booster malfunction occurs and prevents launch within the window, the mission must be postponed until the next suitable alignment.

When I was part of a crew bound for the ISS on shuttle *Atlantis*, the launch window was just five minutes long. A technical glitch in the final seconds of the countdown caused a nerve-racking two-minute delay, but we finally received our GO for launch—we made the window!

5. What do astronauts eat right before launch?

There aren't any limitations on what we can eat. We try to eat our favorite foods because we know space food can never match the variety, texture, and flavors of our favorite dishes on Earth. So when NASA dieticians asked us what we wanted to eat during our week-long medical quarantine, we asked for favorites like pizza, burgers, lasagna, and Tex-Mex.

When asked to choose the breakfast foods I wanted on the last few mornings before liftoff, I ticked off the healthy things I usually had at breakfast: cereal, yogurt, orange juice, and coffee. Crewmate Kevin Chilton interrupted me, saying, "Tom, forget about that healthy stuff—this is the time to let loose!" So I indulged in chocolate-flavored cereal every morning in quarantine and was glad I did! On launch morning, knowing it would be some six hours before I could eat in space, I downed a ham and cheese omelet, toast, hash browns, and orange juice. I left Earth full—and content.

6. Were you worried during the countdown?

My biggest worry was not the risk of flying on the shuttle, but whether I had prepared adequately for my duties in orbit. Expert technicians had prepared the shuttle for flight—I trusted that our rocket was ready. What I experienced instead was the same anxious, light feeling in the stomach that you get before speaking in public or taking a big test in school. Why? I didn't want to be the one who let down my crewmates or my colleagues in Mission Control. I realized, though, that the time for studying was over. It was time to do the job for which I'd been trained.

Still, there is no denying a sense of physical relief at getting safely to orbit. When my pilot Kevin Chilton floated down to the shuttle mid-deck an hour after launch to get out of his suit, he grabbed me by the shoulders and exclaimed with a big grin, "Hey, Tom! We're in space—and we're alive!"

7. What happens if there is a problem in the final seconds before liftoff?

One of the tensest moments in my spaceflight career occurred on board the *Endeavour* during mission STS-68, when one of our main engine turbo-pumps overheated seconds before liftoff. The orbiter flight computers detected the failure just 1.9 seconds before booster ignition, and shut down the main engines.

Instead of the massive bang of booster ignition, we heard the engine noise die away and the simultaneous blare of the master alarm. My crewmates and I unstrapped from our seats and waited tensely for a possible evacuation order. For several minutes we could feel *Endeavour* swaying on the launch pad from that initial kick from the main engines.

Launch controllers quickly checked for fire or explosive propellants in the engine compartment. We sat tight while they worked to return the shuttle to a safe condition. The launch team executed their procedures perfectly, ensuring the safety of both the crew and the shuttle.

8. Is the ride to space painful?

The ride to space is not exactly relaxing, but the experience is so exhilarating that it outweighs any sense of discomfort. For example, the *Soyuz* hits a peak acceleration of about 4 g's during its nearly nine-minute ride to orbit. (The everyday acceleration due to gravity that we experience on Earth's surface is called "1 g," and rocket acceleration levels are described as multiples of this familiar feeling of normal weight. So, an acceleration equal to four times the acceleration at Earth's surface is called "4 g's." At 4 g's, you'd feel four times your normal weight.)

The shuttle reached a peak acceleration of 3 g's for about the last minute of its ride to orbit. Designers held the shuttle's launch acceleration to 3 g's so its structural limits would not be exceeded. At 3 g's, my body felt like it weighed about 500 pounds (227 kilograms); it's hard to breathe, and challenging to lift and direct your hands accurately (to flip a switch, for example). But it wasn't painful, just a feeling of strong and steady pressure. We always joked to each other on the ride "uphill" that "that gorilla is sitting on my chest again!"

The Saturn V Moon rocket's first stage subjected its Apollo crews to 4 g's, and Mercury and Gemini astronauts endured about 7 g's during the second stage of their rides to orbit. Astronauts are prepared for the acceleration and there is nothing to do but bear it, knowing engine shutdown and free fall are just seconds away.

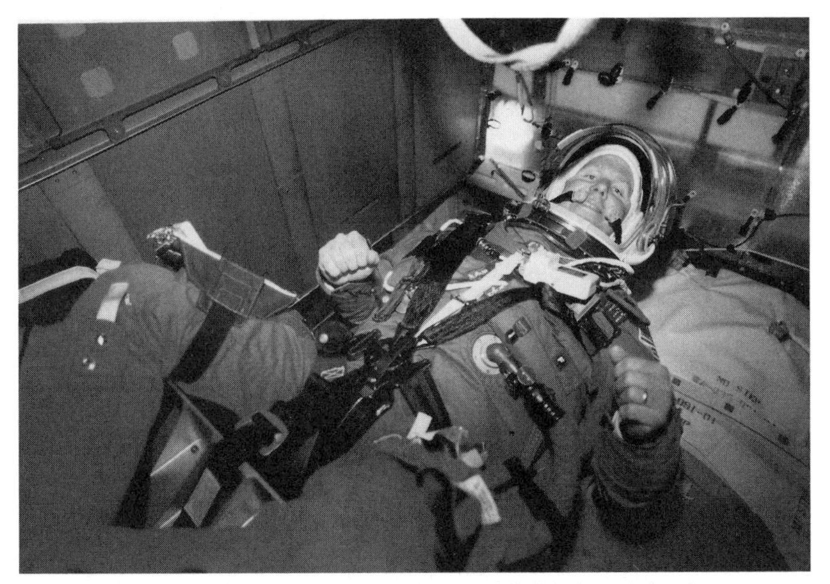

The author strapped in for launch on his second flight aboard shuttle Endeavour. (NASA)

9. What was your shuttle launch experience like?

When the main engines ignited six seconds before liftoff, the entire orbiter rattled and shuddered like a skyscraper in an earthquake. At T-minus-zero, the solid rocket boosters ignited, giving me a massive kick in the back as they blasted our ship off the pad. The pounding exhaust from the twin boosters shook us continually as we accelerated at 2.5 g's, ripping through the lower atmosphere under seven million pounds of thrust.

Two minutes after liftoff the empty boosters fell away with a giant bang, bathing our flight deck in a momentary flash of separation motor exhaust. The three main engines still ran at over a million pounds of thrust—but with almost no vibration—pushing us upward with a comfortable 1-g acceleration.

As the external tank fed those engines and grew ever-lighter, though, the shuttle gradually accelerated to 3 g's and kept us there for the final minute of the ride to orbit. That sustained acceleration was an attention-getter. It felt as if two of my friends were standing on my chest and wouldn't get off! At main engine cutoff, thrust dropped to zero in just a half-second, the pressure on my body vanished, and we were afloat under our straps in free fall at last.

10. How long does it take to reach Earth orbit?

It takes anywhere from 5.6 to 11.7 minutes, depending on the rocket used. The Soyuz rocket reaches orbit in 8 minutes, 45 seconds. The Falcon 9 commercial space booster can lift its *Dragon* cargo capsule to orbit in just over 9 minutes. My space shuttle ascents took about 8 minutes, 30 seconds. The Apollo-Saturn V rocket propelled its astronauts into orbit in 11.7 minutes. The 1960s Gemini astronauts rode their Titan II rocket, an Air Force intercontinental ballistic missile, into orbit in just 5 minutes, 36 seconds.

The ascent times vary because rocket designers consider engine thrust, the rocket's structural strength, and the acceleration limits of the spacecraft and crew to plan a safe path to orbit. Keeping acceleration levels low, for example, means it takes longer to get free of the drag of the atmosphere and to reach orbital velocity.

11. How do you know when you've reached space and become an official astronaut?

An astronaut is a space traveler who is trained to be a professional member of a spaceship crew. However, the definition of "space" has changed over time. In the early 1960s, the U.S. Air Force awarded astronaut wings to its X-15 rocket plane pilots who climbed 50 miles (80 kilometers) or more above Earth. Today, the Federation Aeronautique Internationale sets 62 miles (100 kilometers) above Earth as the arbitrary boundary of space. That's where a NASA astronaut now earns his or her astronaut wings on the way to orbit.

On my first launch, the threshold was still 50 miles. I remember our commander, Air Force Colonel Sid Gutierrez, calling out on the intercom at 50 miles saying, "Congratulations, Tom. You're now an astronaut!" When we returned to Earth, Sid officially presented me with my gold astronaut lapel pin—for me, a dream come true.

12. Do your ears pop on the way to orbit?

On most spaceships, the cabin is sealed before liftoff, so the pressure inside remains the same as it is on the ground. The life support system maintains that internal pressure (usually sea level pressure) all the way to orbit. Because there's no cabin pressure change, your ears do not pop.

Your ears pop in a climbing airplane because the pressure outside the plane falls as altitude increases. In a climb, the air pressure inside the plane decreases, too, to minimize the structural load on the cabin shell and to reduce the engine power needed to run the pressurization system. That system lets the cabin pressure decrease to a level of about 10.9 pounds per square inch, equal to the pressure at about 8,000 feet (2438 meters) above sea level, and maintains that pressure to cruise altitude. The popping sensation you feel during the climb is the higher pressure air inside your inner ear escaping to your throat via your Eustachian tubes, trying to match the pressure of the thinning air outside your skull.

13. Is the launch pad damaged by liftoff?

Shuttle launch pads at the Kennedy Space Center always sustained some damage from the 7 million pounds of rocket thrust unleashed at liftoff. Paint on the launcher deck was scorched and peeled, steel plates cracked under the acoustic pounding of the booster exhaust, and sometimes firebricks from the flame trench (the deep slot in the pad beneath the engines) were dislodged and hurled hundreds of yards out to the perimeter fence.

To cool the launch pad and muffle the intense sound, water nozzles flooded each pad with 900,000 gallons of water per minute, piped from an adjacent water tower. A red plastic "water mattress" was also slung under each of the two booster nozzles to absorb some of the intense blast of ignition. When visiting the pad after a launch, I'd see that shredded red plastic embedded in the chain link fence 500 yards from the launch pad.

Similar techniques will be used to protect the mobile launch platform of the powerful new Space Launch System (SLS) rocket that will launch astronauts on deep-space journeys.

14. Why do we need rockets to reach space?

We use rockets because they are powerful enough to reach the high speeds needed to get into space and stay there. A spacecraft must travel at a speed of about 17,500 miles per hour (28,175 kilometers per hour) to stay in orbit around Earth. That's eight times faster than a rifle bullet! Spacecraft must travel even faster—25,200 miles per hour (40,555 kilometers per hour)—to escape Earth's gravitational pull for trips to the Moon, an asteroid, or another planet.

We can't use jet engines in space because they need oxygen from the surrounding air to burn their fuel, and there is no oxygen in the near-vacuum of space above Earth's atmosphere. Rocket engines carry their own oxygen along with fuel, and together these powerful chemical propellants generate the large amounts of thrust needed to accelerate a spacecraft against the force of gravity. This ability to operate outside the atmosphere makes rockets ideal for sending spacecraft to explore space and the solar system.

15. Which rockets take people to space today?

Many nations launch spacecraft into orbit or across the solar system, but only a few carry people. Russia launches its cosmonauts (and some NASA astronauts) to the International Space Station on the Soyuz rocket. China launches its astronauts on the Long March 2F rocket. NASA is also contracting with commercial companies to use the Atlas V and Falcon 9 rockets to launch its own and partner astronauts to the ISS.

NASA is building the Space Launch System (SLS) rocket to take astronauts to deep space. The SLS will carry *Orion*, a deep-space crew capsule. For brief trips to space, suborbital tourism companies will launch passengers on rocket-powered space planes or capsules like *SpaceShipTwo, Lynx*, or *New Shepard.*

16. Are there other ways to get to space?

Some spacecraft designers hope to use a supersonic cruise ramjet engine, called a scramjet, to accelerate a winged vehicle high into the atmosphere to near-orbital speeds. Because it burns its fuel using oxygen in the air, a scramjet doesn't have to carry large oxygen tanks and is thus lighter than a typical rocket. Close to orbital speed, small rockets would give the last little burst of thrust needed to reach orbit. A British company hopes to use scramjets for its *Skylon* space plane.

There are other, more exotic, proposals for reaching orbit. One would use powerful electromagnets to accelerate a ship to high speed on a long railway track. Leaving the inclined track, the spacecraft would coast upward and then fire small rockets to get the speed necessary to reach orbit. Another method would use a large balloon to carry a rocket high into the atmosphere. Once there, the rocket would drop free and fire itself into orbit. This method might work for small satellites, but rockets carrying astronauts would likely be too heavy for a balloon.

17. Do you fire your spaceship's engines to stay in orbit?

Once in orbit, a spacecraft doesn't need to use its engines to stay there—it just coasts. It continues to coast in an elliptical orbit around Earth until some other force pulls it back to the ground. One such force is atmospheric drag, caused by the ship running into the few oxygen or nitrogen atoms in the very thin upper atmosphere. Drag causes the ship to lose energy and altitude. Eventually the ship drops into the upper atmosphere and begins an unwanted reentry. To overcome atmospheric drag, the International Space Station fires its thrusters every few months to restore its speed and altitude.

18. What happens if something goes wrong during launch?

Crewed spacecraft today come with escape systems to ensure that astronauts can survive a launch emergency and return safely to Earth. For example, the Russian *Soyuz* is topped with an escape rocket, which is triggered automatically during ascent if the rocket loses control, loses thrust, or breaks apart.

During a launch pad fire in September 1983, *Soyuz* launch controllers could not activate the escape system because flames had burned through the communications cables. The crew's manual abort switch inside the spacecraft also failed to function. As flames engulfed the rocket, controllers finally were able to activate the escape system with a radio command, successfully rocketing the cosmonauts' descent module to a safe parachute landing nearby.

Space taxis like *Crew Dragon* and the *CST-100 Starliner* have liquid-fueled "pusher" rockets at their base or sides, capable of propelling them free of a malfunctioning rocket. The *Orion* spacecraft is equipped on its nose with a solid-fuel launch abort motor, powerful enough to pull the craft away from its failing Space Launch System rocket. Parachutes would then lower the capsule to a safe splashdown.

Crew Dragon rockets off its Cape Canaveral launch pad in a test of its launch abort system. (SpaceX)

19. Why do you wear a spacesuit during launch?

Spacesuits protect the crewmembers if there is a sudden loss of pressure in the cabin from a leak or a puncture. During a rapid cabin decompression, an astronaut without a spacesuit would lose consciousness in just a few seconds. The suit allows you to keep working to fix the leak or fly the ship safely back to Earth.

The orange Advanced Crew Escape Suit (ACES) that I wore during shuttle launches also would have protected me from strong winds, extremely cold air temperatures, and cold ocean waters if I had to bail out and make an emergency parachute landing. The spacesuit and harness contained a life raft, oxygen, water survival gear, and marker lights so rescue forces could more easily locate a downed astronaut.

20. Why was the shuttle spacesuit orange?

The "international orange" color was chosen for safety reasons. It was highly visible to rescue crews searching for astronauts floating on a blue, green, or gray ocean. To provide even better visibility, the pressure suit's white helmet was covered with brilliantly reflective white tape. On the upper part of each arm, astronauts also wore chemical light sticks that showed up brightly in a helicopter pilot's night-vision goggles. Future astronauts launching near or landing on the water may also wear orange suits for rescue purposes.

21. Which NASA missions experienced a serious failure on the way to orbit?

The *Apollo 13* crew was riding the S-II second stage of their Saturn V rocket on April 11, 1970, when one of the engines on that stage shut down due to excessive vibration. The remaining engines on the second and third stages got *Apollo 13* into the proper orbit to proceed toward the Moon.

In July 1985, the astronauts of shuttle *Challenger* on mission STS-51F were nearly six minutes into their ascent when their center main engine shut down due to multiple sensor failures. Quick action by flight controllers in Houston prevented another sensor-caused engine shutdown, and *Challenger* executed an abort to a lower-than-planned orbit and completed a successful mission.

Just six months later, shuttle *Challenger* (mission STS-51L in January 1986) was destroyed by a flawed solid rocket motor that severed its connection to the external propellant tank and crashed into the tank and orbiter. The breakup of the orbiter sealed the fate of *Challenger's* crew, which perished when their crew cabin struck the ocean surface off Cape Canaveral.

When shuttle *Columbia* launched in January 2003, its left wing was struck and damaged by a chunk of external tank insulation. During reentry on February 1, 2003, the damaged heat shield panel on the left wing allowed hot plasma to enter and destroy the wing structure. *Columbia* broke up at about 12,000 miles per hour (19,400 kilometers per hour) over Texas, and the crew did not survive.

22. Did the shuttle have a launch escape system?

The shuttle had a very limited launch escape system, which was installed after the *Challenger* accident in 1986. Crewmembers could jettison the side hatch and jump free of the orbiter using parachutes. However, astronauts could only bail out if the orbiter managed to get free of its solid-fuel rockets and fuel tank and get into a stable glide, which was unlikely during a serious launch failure.

After both the *Challenger* and *Columbia* shuttle accidents, designers considered adding an escape capsule to the shuttle cabin, but NASA decided that option was too expensive. The agency opted instead to improve the shuttle's overall launch and reentry reliability, but this approach did not save the *Columbia* crew. The lack of a good escape system was an important factor in the decision to retire the shuttle in 2011.

23. What are the differences between a suborbital and an orbital flight?

A rocket that doesn't get to a speed of 17,500 miles per hour (28,175 kilometers per hour) will fall back to Earth before completing an orbit of the planet. This is called a suborbital flight. But the spacecraft may still leave the atmosphere, and humans aboard will see the black sky of space and experience the weightless sensation of free fall for a few minutes before reentering the atmosphere. Astronauts Alan Shepard and Gus Grissom flew suborbital missions on America's first two spaceflights in 1961.

When a spacecraft reaches the correct altitude with a speed of 17,500 miles per hour, traveling roughly parallel to the Earth's surface, it can maintain an orbit around the Earth even after the engines shut down. This is called orbital flight, and it continues until atmospheric drag pulls the spacecraft down, or until small braking rockets are fired to drop the craft back into the atmosphere.

24. Was each of your launches as exciting as the first?

There's nothing like your first trip into space. Part of the excitement is the anticipation of the unknown, and experiencing all the physical and emotional exhilaration of launch for the first time. On my next three missions, I was still thrilled by the physical sensations of launch, but I was able to focus more on the details of the orbiter's performance and my own reactions to acceleration and free fall.

Although the "gee whiz!" factor might have been slightly less on my later launches, the physical thrill was every bit as amazing. It's marvelous to sit at the top of a rocket and experience firsthand a process that accelerates your body to eight times the speed of a rifle bullet and puts you precisely into a near-circular orbit of the Earth.

25. What did you think about before launch?

For each of my six launch countdowns, I was strapped into my seat about three hours before liftoff. My crew monitored the countdown as the launch controllers prepared to rocket us into space. During this time, we were paying careful attention to those checklists and the state of the spacecraft.

But there were many minutes when we had little to do but listen. I thought about our mission and the work I was to do in orbit. We cracked jokes, shared past launch stories, and tried to keep the mood light. I often prayed, asking God for safety and success on the mission, for my family's future safety and happiness, and for help in doing my job once I got to space.

26. What were your first impressions of spaceflight?

I was very excited to experience my first launch and orbit of the Earth. After the tremendous vibration and acceleration of launch, I next experienced the strange and exhilarating sensation of free fall. A frenzy of work in the orbiter *Endeavour's* middeck followed as I helped convert our rocket ship into an Earth-observing science laboratory.

After more than an hour of work in the cabin, I was able to peer out the orbiter's side hatch window for a few minutes and watch an exquisite sunrise. I was in awe as the morning light washed over the lovely Earth below—such astonishing colors! A surprise for me after reaching orbit was the intense closeness I felt toward my crewmates. These five had all helped me get to orbit, and each day they did their best to help with my work and share their own exhilaration at being in space.

27. Were you quarantined after your mission?

I was not. When the first three Apollo crews returned from landing on the Moon, they were quarantined for three weeks. Scientists were concerned that they were carrying Moon germs that might infect Earth. No germs were found, and so the last three Apollo crews were released immediately after landing, as are returning Earth-orbiting crews today.

The first crews to return from the surface of Mars, where hardy micro-organisms can theoretically survive, will face some period of quarantine. They will likely begin it during the months-long voyage back to Earth, then remain in quarantine until surface samples from Mars have been analyzed and found not to harbor any harmful organisms.

28. Who were the first astronauts to reach space?

Yuri Gagarin, a Russian representing the Soviet Union, reached space and orbited the Earth once on April 12, 1961. Alan Shepard became the first American in space when he flew a 15-minute suborbital mission on May 5, 1961. American Gus Grissom flew another short suborbital hop just over 100 miles (167 kilometers) up on July 21, 1961. Cosmonaut Gherman Titov orbited Earth for over a day on Aug. 6, 1961, and American astronaut John Glenn circled the planet three times on Feb. 20, 1962.

29. How many people have gone into orbit?

As of 2015, 542 humans had orbited the Earth. The United States has launched 344 people into space, the most of any nation. Twenty-four have circled the Moon, and 12 have walked on the lunar surface. The latest space traveler counts can be found on the website Astronaut/Cosmonaut Statistics: www.worldspaceflight.com/bios/stats.php.

30. How much power is needed to lift the Space Launch System rocket off the ground?

NASA's Space Launch System (SLS), taller than the Statue of Liberty, will carry the *Orion* deep-space crew vehicle or heavy cargoes into orbit. It will produce 8.4 million pounds of thrust at liftoff to hurl its 5.5-million-pound weight skyward, more thrust than the Apollo-Saturn V Moon rocket. That's more than the thrust from 35 Boeing 747 jumbo jets, and matches the power produced by 160,000 Corvette engines or 13,400 locomotives.

The SLS will be able to haul more than 77 tons (70 metric tons) to orbit. The launch thrust of this rocket is 10 percent more than the Apollo-Saturn V rocket's 7.5 million pounds and 20 percent more than the shuttle's 7 million pounds of thrust.

31. What will space tourists experience during their suborbital flights?

Space tourists will experience much of the same vibration, noise, and acceleration during launch that I did. Passengers aboard the *SpaceShipTwo* and *Lynx* space planes and the *New Shepard* rocket will experience about 4 g's of acceleration on their brief ride to the edge of space.

However, they will enjoy only about 10 minutes of free fall during their short, cannonball-style rides to 62 miles (100 kilometers) up.

Passengers will have time to steal a glance at the black daytime sky, see the Earth's curved horizon and gauzy atmosphere, stare down at the sweeping, pastel landscapes below, and float ever so briefly in their small cabin space. All too soon, they will need to strap in again for the plunge through the atmosphere to a safe landing.

CHAPTER IV

Ships for Space

NASA's Space Launch System rocket, built to carry heavy loads to deep space, lifts off in this artist's concept. (NASA)

1. How can a rocket work without air to push against?

Sir Isaac Newton discovered three mathematical laws that govern how objects move in our universe. His third law says that for every action, there is an equal and opposite reaction. In other words, throw something one way, and you will be pushed in the opposite direction.

That's how a rocket works. The "action" is the force pushing its hot exhaust gases down out of the engine's nozzle, and the "reaction" is the force that pushes the rocket in the opposite direction, up toward the sky. A rocket requires nothing to push against to generate that opposite force; it must only push out hot gas at high speed in one direction to generate a push in the other. Thus, rockets can work in a vacuum.

A good analogy is the kick you feel from a garden hose when it shoots out a stream of water. Another is what happens to an inflated toy balloon when you release its neck; it jets away in the opposite direction of the escaping air.

2. What did you see and feel when the engines fired in space?

We are used to seeing bright flames and fiery rocket exhaust when a rocket lifts off from Earth, but astronauts see very little of that dramatic action in the vacuum of space.

In orbit on board the shuttle, we seldom fired anything but small thruster rockets, which usually blasted for only a fraction of a second. These were used to pivot our spacecraft to point in a particular direction and to adjust the height or shape of our orbit. The shuttle had 44 rocket thrusters: 38 large ones that each produced 870 pounds of thrust and six small ones that each produced just 24 pounds of force.

During the daytime, these flashes of rocket exhaust were nearly invisible. At night, we'd see a momentary glowing jet of gas from a thruster nozzle. The large rear thrusters would send a slight thump through the shuttle's structure, while the close-by forward thrusters, just in front of the crew cabin, fired with a solid WHUMP!

The best rocket show I saw in orbit was just after we started a nighttime reentry of the shuttle *Atlantis* during mission STS-98. To burn off the heavy, excess rocket fuel in our forward thruster system (and thus balance our weight for landing), we ignited the 870-pound-thrust engines for almost a minute. This sent a triple plume of brilliant, yellow-white exhaust jetting upward ten feet in front of our windows: an amazing sight!

3. Does the International Space Station need rocket engines to stay in orbit?

Most of the time the ISS simply coasts around Earth, with engines silent, at an average altitude of 248 miles (400 kilometers). However, the station's large solar panels and modules are constantly running into tiny air molecules that slow it down. This results in a small but steady decrease in the station's altitude. Without firing its engines every so often, the ISS would be dragged lower and would eventually reenter the Earth's atmosphere and burn up. It takes four tons of rocket fuel each year to keep the ISS in space.

The space shuttle's twin orbital maneuvering system engines fire to maneuver the spacecraft. (NASA)

4. How does the International Space Station receive its supplies?

The ISS crews receive supplies such as water, food, spare parts, and new science experiments via unmanned cargo ships. NASA contracts with two companies, Orbital/ATK and SpaceX, to send robotic cargo craft to the station about six times per year. Russia also sends robotic freighters to the ISS about once every three months.

Only SpaceX's *Dragon* cargo ship is able to return large amounts of obsolete equipment and scientific samples to Earth. *Dragon* uses a heat shield, which keeps it from burning up on the way down, and a parachute to land in the ocean off California. The other cargo ships are unloaded, then packed with trash and cast off to incinerate as they enter Earth's atmosphere.

5. How high is the International Space Station?

The ISS is usually about 248 miles (400 kilometers) above Earth. The altitude varies slightly as drag from Earth's atmosphere pulls the station lower, and as periodic "reboosts" from rocket thrusters raise it higher.

6. How big is the International Space Station?

It's about the same size as a football field including both end zones. It measures 357 feet (109 meters) wide, and its solar array wingspan from tip to tip is 240 feet (73 meters), longer than the wings of a Boeing 777 200/300 model, which measure 212 feet (nearly 67 meters). The ISS weighs almost one million pounds (approximately 925,000 pounds, or 420,000 kilograms). That's the equivalent of more than 320 automobiles.

The ISS is almost four times as large as the Russian space station *Mir* and about five times as large as the 1970s U.S. *Skylab* station. The station has the interior living space of a six-bedroom house.

7. Can you see the International Space Station in the sky?

Yes, the space station is one of the most delightful sights seen in the sky—a brilliant star-like object with six people living aboard it. You can see it without a telescope or binoculars, despite its being 248 miles

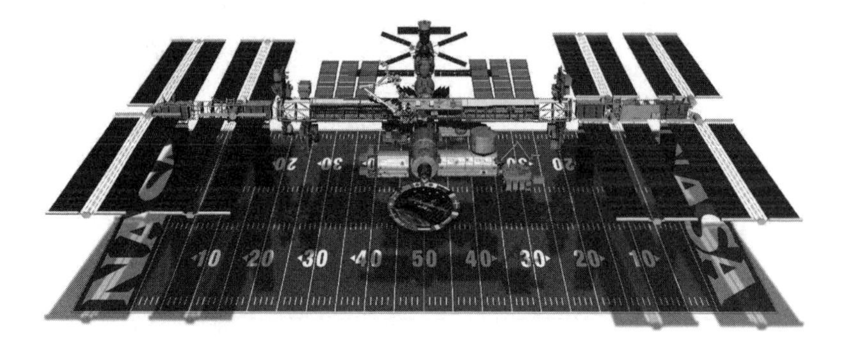

The ISS would cover most of a football field back on Earth. (NASA)

above the Earth's surface. Its vast solar arrays and shiny aluminum structure catch the rays of the Sun and reflect them to our eyes below.

The station can be seen just before sunrise or right after sunset, when the orbiting outpost is still bathed in sunlight and the Earth below is in twilight or darkness. The ISS appears as a brilliant, slowly moving star—there is no mistaking it. But if you spot a star that blinks regularly, it's an airplane! Find out where and when to see the ISS at NASA's Spot the Station website: spotthestation.nasa.gov.

8. When was the International Space Station built?

The ISS was built piece-by-piece between 1998 and 2011. Construction began in December 1998 when the STS-88 space shuttle crew brought up the U.S. *Unity* module and attached it to the Russian FGB cargo module, which had been launched into orbit earlier. The space shuttle delivered NASA's last major module, the Permanent Multipurpose Module, *Leonardo*, to the ISS in 2011.

Fifteen pressurized modules make up the ISS living and work space. The U.S. contributed seven modules: *Destiny, Unity, Quest, Tranquility, Harmony, Cupola*, and *Leonardo*. Russia has launched five modules: *Zarya, Zvezda, Pirs, Poisk* and *Rassvet*. Japan built two modules: the *Kibo* lab and its logistics module, and the European Space Agency

The International Space Station as completed in 2011. (NASA)

contributed the *Columbus* science lab. Cargo ships still deliver new experiment packages for installation on the outpost's exterior. Russia plans to add a multipurpose laboratory module and a solar power module to the outpost.

9. How many missions did it take to assemble the International Space Station?

The ISS is made up of 40 major pieces, including the modules where the astronauts live and conduct experiments. It took 37 space shuttle missions and three Russian launches to bring them all up into space along with the robot arm, the solar arrays, and the bridge-like trusses that support them.

Another 113 launches—mostly Russian—between 1998 and 2014 brought research gear, crews, and cargo to the outpost. Additional launches continue at the rate of about 10 – 12 per year. To find out more about launches to the ISS, go to NASA's International Space Station Facts and Figures website:
www.nasa.gov/mission_pages/station/main/onthestation/

10. What is the expected lifetime of the current International Space Station?

NASA wants to keep the outpost running at least until 2024. Engineers are tracking the wear and tear on the station, and it appears to be aging gracefully.

Sometime in the mid-2020s, small cracks from metal fatigue may begin to form in some sections, and systems onboard may start to break down. For example, we might detect fluid leaks, corrosion damage, degraded solar panels, or worn-out motors.

NASA and its partners will decide if they can salvage any working parts from the station, and possibly use them in building a new outpost. The parts that are no longer usable or safe will either be guided to a splashdown in an empty ocean area or rocketed higher into a safe disposal orbit.

11. How does NASA get its astronauts to the International Space Station?

Russia, a major ISS partner, sells seats on its Soyuz rocket to NASA. The U.S. space agency buys about six seats annually, which are used by American astronauts and partner fliers from Japan, Canada, and the European Space Agency. Russia has provided regular crew transport since 2010 after the shuttle delivered the last few major parts of the ISS. NASA will continue to buy these services until American space taxis begin flights to ISS, probably in 2018.

12. How long will NASA keep flying astronauts on the Russian Soyuz?

Once American space taxi companies SpaceX and Boeing complete successful test flights of the *Crew Dragon* and *CST-100 Starliner* spacecraft, they will begin regular transportation to the International Space Station. But NASA will still fly some astronauts on the Russian *Soyuz,* because it always wants to have at least one U.S. station crewmember qualified to use the *Soyuz* as an emergency lifeboat. That way, if American taxi services are temporarily halted, a NASA astronaut can remain at the station with his or her Russian partners and rely on the

Russia's Soyuz spacecraft. Astronauts ride in the descent module, at center. (NASA)

Soyuz for emergency transportation. For the same reason, some Russian cosmonauts will ride the *Crew Dragon* or *CST-100 Starliner* to and from the ISS.

13. How many countries can send people into space today?

Russia and China are the only countries currently operating their own rockets and manned spacecraft. The U.S. retired its shuttles in 2011, but plans to fly commercial spacecraft that will carry NASA astronauts to the space station. U.S. astronauts should resume flights from Cape Canaveral sometime in 2018.

14. Was the International Space Station the first space station?

Several space stations came and went before the ISS. In 1971, the Soviet Union launched the first space outpost, *Salyut 1*. Then the U.S. launched the *Skylab* space station, which operated between 1973 and 1974 with three visiting crews. The Soviets followed up with *Salyut 3*,

Salyut 4, and *Salyut 5* between 1974 and 1977. Next came *Salyut 6* and *Salyut 7,* which were in space for a long time, from 1977 to 1991.

Russia's *Mir* space station was periodically occupied between 1986 and 2001, and set human endurance records for cosmonaut stays in orbit. *Mir* also taught Russia, the U.S., and its partners how to collaborate in planning and building the ISS.

15. How long have astronauts been living on the space station?

The first astronaut crew—Bill Shepherd, Yuri Gidzenko, and Sergei Krikalev—to live aboard the International Space Station docked there on November 2, 2000. Since then, the station has been occupied continuously by a series of astronaut and cosmonaut crews. These stays on the ISS are called expeditions. You'll find a running total of how long humans have lived off the planet at NASA's International Space Station website: nasa.gov/mission_pages/station/main/index.html.

16. Why couldn't the shuttle go to other planets and the Moon?

The space shuttle was a big vehicle, weighing over 100 tons. To reach the Moon or other planets, the shuttle would have had to accelerate to a speed of 7 miles per second (11 kilometers per second) in order to escape Earth's gravitational pull. The shuttle simply did not carry enough fuel to reach such a speed.

The shuttle was never designed to leave Earth orbit. Instead, it was designed to reach and maneuver in low Earth orbit, where it could use its limited fuel supply to launch and retrieve satellites, assemble a space station, repair existing satellites, launch probes to the planets, and serve as a science platform. The shuttle was the most versatile spacecraft ever built, with wings and wheels that enabled it to land on a runway and be reused many times.

17. Why was the space shuttle retired in 2011?

The shuttle gave way to new goals for deep space exploration, where it could not venture. After the destruction of shuttle *Columbia* in 2003, President George W. Bush ordered the retirement of the shuttle orbiter fleet by 2010, once the construction of the International Space Station was completed. The plan at the time was to use the new *Orion* spacecraft to get astronauts to the International Space Station by 2012, and then take them to the Moon by 2019. A lunar outpost and then Mars were NASA's deep-space goals back then.

President Obama agreed with the plan to retire the shuttle, but decided to skip the Moon and instead send astronauts to a nearby asteroid as a training location for reaching Mars in the 2030s. The last shuttle was retired in 2011. But due to tight budgets, construction of the deep-space *Orion* spacecraft and its new rocket booster, the Space Launch System, was delayed. The earliest that *Orion* will carry astronaut crews will be in the early 2020s.

18. Where are the space shuttle orbiters exhibited?

The space shuttle orbiter *Discovery* is now on exhibit at the National Air and Space Museum's Udvar-Hazy Center near Dulles International Airport in Virginia. *Endeavour* is at the California Science Center

The author with space shuttle orbiter Atlantis at the Kennedy Space Center Visitor Complex. (Peter W. Cross)

in Los Angeles, California. *Atlantis* is at the Kennedy Space Center Visitor Complex in Florida. *Enterprise,* the prototype shuttle that flew only in atmospheric landing tests, is at the Intrepid Sea, Air, and Space Museum in New York City.

19. Which shuttle orbiter was your favorite?

I liked *every* orbiter I flew. *Endeavour* was my first—always a sentimental choice. *Columbia* was the very first shuttle to fly, in 1981, and I felt privileged to serve aboard this historic ship. *Atlantis* took me to the space station and gave me three memorable spacewalks. Visiting *Atlantis* in Florida and *Endeavour* in California is always an emotional experience for me. These ships are like old friends. I have very special memories of *Columbia*, flown by so many heroes and friends before that orbiter was lost in 2003. Don't miss the opportunity to see *Enterprise, Discovery, Atlantis*, and *Endeavour* if you are near one during your travels.

20. How fast did you go in space?

The fastest I traveled in space was roughly 17,694 miles per hour (28,475 kilometers per hour). At the International Space Station on my last mission, I was traveling 17,100 miles per hour (27,475 kilometers per hour). This is approximately 25 times the speed of sound, also known as Mach 25! I got a special patch from NASA after my first trip at that speed.

At this speed, the ISS completes an orbit of the Earth once every 92.7 minutes. Space station crews cross the vast expanse of the Pacific Ocean in a little over 28 minutes, and see a sunrise or sunset about every 45 minutes.

By comparison, the fastest jet aircraft ever flown was the Lockheed SR-71 Blackbird reconnaissance aircraft, which cruised at Mach 3.

21. What are the highest speeds astronauts and spaceships have achieved in space?

The highest speed achieved by humans was recorded by the *Apollo 10* astronauts in May 1969, when they hit 24,791 miles per hour (40,161

My crew with Columbia after landing from my third shuttle mission, STS-80. (NASA)

NASA astronauts who flew the space shuttle received this "Mach 25" patch after flying 25 times the speed of sound. (NASA)

kilometers per hour) on their return from the Moon, just before reentering the Earth's atmosphere.

The fastest man-made machine to leave Earth was the *New Horizons* spacecraft, which flew by Pluto in July 2015. When it left Earth in 2006 it was traveling at 36,000 miles per hour (59,000 kilometers per hour).

The fastest speed ever achieved by a spacecraft was reached by the *Helios I* and *Helios II* solar probes, launched in 1974 and 1976. Plunging closer to the Sun than Mercury, the innermost planet, the two probes were accelerated by the Sun's gravity to speeds over 150,000 miles per hour (250,000 kilometers per hour).

22. What new rockets is NASA building?

To carry astronauts to the International Space Station, NASA is hiring commercial companies to build "space taxis," like *Crew Dragon* and the *CST-100 Starliner*. They will be launched into space on rockets like the Falcon 9 and Atlas V, also built and operated by commercial companies. These taxis will carry American astronauts to low Earth orbit for NASA through the 2020s.

To take astronauts into deep space, NASA is testing the Space Launch System (SLS) rocket. It's a two-stage rocket with a pair of giant, solid-fuel boosters attached to each side. SLS can carry three astronauts around or past the Moon, and lift 77 tons (70 metric tons) of cargo to low Earth orbit. Its first test flight in 2018 will carry an *Orion* spacecraft (without a crew) around the Moon.

23. How does the Space Launch System (SLS) compare to the Saturn V Moon rocket?

The first SLS rocket will stand 322 feet (98 meters) tall. The Saturn V rocket, which took astronauts to the Moon, was 363 feet (110 meters) tall. The next-generation SLS rocket, which will carry heavy cargo, will be even taller—384 feet (117 meters)—some 20 feet taller than the Apollo-Saturn V Moon rocket. By comparison, the space shuttle stood 184 feet (56 meters) high, and the Statue of Liberty rises 305 feet (93 meters) from ground level.

24. Will the Space Launch System be more powerful than the space shuttle?

Yes. The first SLS rocket will produce 8.4 million pounds of thrust (3.8 million kilograms) at liftoff, which compares to the Saturn V rocket's 7.5 million pounds, and the shuttle's 7 million. The SLS will lift 70 metric tons to low Earth orbit, compared to the shuttle's payload capacity of over 24 metric tons. This 70-metric-ton payload carried into orbit will include another rocket stage designed to carry cargo or an *Orion* spacecraft into deep space. The next-generation cargo version of SLS will produce 9.2 million pounds of thrust and propel 130 metric tons into orbit.

25. What future space stations are planned?

China plans to put its own space station, the *Tiangong 2,* in orbit before 2020. Plans include visits by three-astronaut crews lasting up to

Saturn V (left) moon rocket compared to the space launch system set to fly in 2018. (Alex Connor Brown)

20 days. Sometime in the 2020s, China proposes to launch the Tiangong 3, a larger outpost that will fly for as long as ten years.

Russia's space agency announced in 2015 that it would orbit a new space station in the 2020s, built in part from some of the newer modules it sent to the International Space Station.

The U.S. firm Bigelow Aerospace has plans for a commercial space station, called Alpha. It would be made up of inflatable living modules 22 feet (6.7 m) across. The outpost would host tourists and could be leased by industrial or commercial clients.

Artist's concept of the Bigelow Alpha space station. (Bigelow Aerospace)

26. What kinds of private spaceships are planned, and how far will they go?

Private, suborbital ships being developed include XCOR's *Lynx*, Virgin Galactic's *SpaceShipTwo*, and Blue Origin's *New Shepard.* All are space planes or capsules designed to reach 62 miles (100 kilometers) on brief, 15- to 30-minute flights to the edge of space.

For transportation to the International Space Station (ISS), two American aerospace companies are building commercial space capsules. Both the SpaceX *Crew Dragon* and Boeing *CST-100 Starliner* are designed to transport crews to the station, remain docked there for up to six months, and then return home. Each will also serve as a lifeboat while docked at the ISS. SpaceX states that its *Crew Dragon* could be upgraded to land on Mars, but that will require a better heat shield, much bigger parachutes, and powerful landing rockets.

Mockup of the Boeing CST-100 Starliner commercial space taxi. (NASA)

27. What ships are planned to take us into deep space?

Orion is NASA's multi-purpose crew vehicle, intended to carry American astronauts beyond low Earth orbit and into deep space in the early 2020s. *Orion* will carry out missions around the Moon, to a captured asteroid in lunar orbit, or to a gravitationally balanced location some 33,000 miles (48,280 kilometers) beyond the Moon called the L2 Lagrange point. At L2, a spacecraft can hold its position over the far side of the Moon while using very little fuel.

Equipped with additional living quarters for deep-space travel and extra engines and propellant tanks, *Orion* can reach destinations several million miles from Earth like near-Earth asteroids. *Orion* will probably be part of a future spacecraft that takes astronauts to Mars.

28. What other kinds of spaceships are needed to explore the Moon, asteroids, or Mars?

We'll need several new spaceships in addition to *Orion* to reach the Moon, 240,000 miles (389,000 kilometers) away, nearby asteroids, or Mars—at least 35 million miles (57 million kilometers) away.

To reach the near-Earth asteroids, *Orion* will need a habitation module to provide elbow room, supplies, and life support systems. An extra propulsion section, containing additional propellant tanks and engines, will be used for space maneuvering. If we want to land on the Moon again or reach the surface of Mars, we'll need to build rocket-powered landers. Engineers think it will be around 2040 before all the technology needed to reach Mars is ready.

Artist's concept of a nuclear-powered, piloted Mars expedition about to leave Earth orbit for the Red Planet. (NASA/John Frassanito & Associates)

29. What future spaceships are planned by other nations?

Both Russia and China plan to build ships to land on the Moon by the late 2020s, but neither has set a specific date for doing so. The Russians plan a new manned spacecraft, the *New Generation Piloted Transport Ship*, to replace the *Soyuz* spacecraft in the 2020s. This spacecraft will also serve as the command ship for a future Russian lunar landing mission. China has announced plans to land astronauts on the Moon by the late 2020s, using its *Shenzhou* spacecraft accompanied by a yet-to-be-built lander.

India also has plans for getting humans into space. In December 2014 it flew an unmanned test version of its Orbital Vehicle, a small piloted capsule.

30. Do we know how to reach the Moon, asteroids, and Mars?

The U.S. Apollo program successfully landed 12 astronauts on the Moon's surface between 1969 and 1972, so we know how to reach our natural satellite.

Near-Earth asteroids routinely approach within a few million miles of our planet. To reach a few of these it will take less fuel than for a roundtrip journey to the Moon. Most near-Earth asteroids are no larger than a mile or so across. They have very little gravity, so a spacecraft can pull up alongside them with no need for a separate lander spacecraft.

Mars is a much bigger challenge. It takes 6 to 9 months to reach Mars. Its atmosphere is too thin for parachutes to land spacecraft safely, and even though its gravity is only one-third that of Earth, it is strong enough to require rockets for final descent. Engineers don't yet know how to build a landing system good enough to get a heavy spacecraft and astronauts safely down to the surface of Mars.

The Crew Dragon commercial space taxi. (NASA)

31. Can tourists visit the space station?

Yes, but it costs a lot of money to do so—more than $50 million per trip. Russia has sold a number of seats on its *Soyuz* spacecraft to allow private space travelers to visit the International Space Station (ISS). These sales help pay for their space program. The first tourist to visit the ISS was Dennis Tito in 2001. More private spaceflight opportunities may open up when U.S. commercial space taxi companies begin launching from Cape Canaveral, Florida.

32. Can I go on a vacation in space?

Boeing or SpaceX, both U.S. companies, promise trips to future space outposts on their new space taxis, but the price has yet to be set.

These ships will not take you to the International Space Station, but to private space stations like the inflatable Bigelow Alpha station, where you'll be able to enjoy a week or more of free time off the planet. Personal space travel will be very expensive at first, but with more companies competing for travelers and cargo the costs will come down, making trips more affordable over time.

CHAPTER V

Surviving in Space

Japanese astronaut Soichi Noguchi works outside the U.S. Destiny lab at the International Space Station. (NASA)

1. How can you tell which way is up while in space?

When living weightless in free fall, *you* choose where you want "up" to be. Your mind will happily accept this arbitrary choice, no matter where Earth is outside the window. Most astronauts choose "up" to be in whatever direction their head is pointing. For this reason, at any given time there can be as many "ups" as there are astronauts.

On my first two shuttle flights, our orbiter *Endeavour* usually flew upside down relative to Earth to accomplish its Earth observing mission. Its cabin ceiling and overhead windows pointed down at the ground, and so did we. Floating upside-down, heads toward Earth, we were better able to observe our science targets below. But we weren't bothered at all by the sight of our beautiful globe hanging over our heads as we hovered near the flight deck's sunroof-style windows. On my fourth mission, while *Atlantis* was docked at the International Space Station, Earth remained in its familiar place, beneath the station's deck. We were equally happy with both situations.

2. What is free fall?

Although free fall and weightlessness describe the same condition, free fall is the more accurate term. Free fall is the special type of motion that happens when the only force acting upon an object is gravity—there are no other forces such as friction, air resistance, tension, or anything pushing or pulling on the object.

Newton's second law of motion and law of gravitation tell us that in free fall all objects will fall with the same acceleration, regardless of their mass (which on Earth we refer to as "weight"). This is the situation in a spacecraft orbiting Earth: the spacecraft and its occupants and all the objects in the vehicle are actually falling around Earth. It's as if you're all in an elevator plunging down a shaft in a tall skyscraper. Inside the elevator, everyone and everything appears to "float," because you're all accelerating at the same rate. Fortunately, astronauts in this special condition of orbital free fall never hit the bottom of the elevator shaft.

3. What causes astronauts to float while in free fall?

Weightlessness is a term sometimes used to describe what you feel when there is nothing touching your body and nothing is pushing or pulling you. The sensation of weightlessness happens any time you are in a state of free fall. When a spacecraft is orbiting the Earth, the astronauts inside are in a state of continuous free fall—they are literally falling around the Earth.

Suppose you go on a drop tower ride at the amusement park. As you fall from a very high tower, both you and your seat are accelerated toward the ground at the same rate by Earth's gravity. Because the chair is accelerating at the same rate as you are, it can't push on your bottom or exert a force on your body. You will feel yourself floating just above the chair—"weightless"—even though gravity is pulling on both of you.

4. What does free fall feel like?

Entering free fall is a remarkable sensation. At engine shutdown, all acceleration pressure on the body instantly vanishes. Your body feels buoyant. Instead of being pushed heavily back into your seat, you rapidly perceive that you're gently floating under your seat straps. The closest thing to experiencing free fall on the ground is the sensation of floating on your back in a warm pool, with nothing touching your body but the water. You feel effortlessly supported and buoyant. When floating like this, I always gaze up at the sky and pretend I'm looking down at Earth's blue oceans. Try it!

5. Why can free fall make you feel sick to your stomach?

Physiologists think that in orbital free fall, the unusual signals coming from the otoliths in your inner ear—tiny organs that control your balance—conflict with what your eyes are telling your brain about which way is down or up. Your confused brain sometimes reacts by emptying your stomach.

About a third of astronauts feel fine when they arrive in free fall, a third feel a bit out of sorts, and a third experience bouts of nausea. Flight surgeons call the condition "space adaptation syndrome," more commonly known as "space sickness." Fortunately, the medicine available for it works very well, curing the nausea quickly.

On my first mission, I suffered from nausea soon after launch, but an injection cured that and quickly restored my appetite. Before my next launch, I took the same medication an hour before liftoff, while sitting on the launch pad, and never felt a twinge of space sickness.

6. How long does it take to adapt to the space environment?

Most astronauts take two to three days to feel at home in free fall conditions. It's not just the outside of your body that is floating around in free fall; everything inside you is, too! This includes your body fluids —blood and lymph—are no longer tugged down into your legs and lower torso by gravity. As fluids migrate to the head, your face looks fuller, you feel flushed, your sinuses get congested, and you may get mild headaches. You will also have to take frequent trips to the bathroom, as your body rids itself of what it considers excess fluid in the upper body. But once the fluid balance in your body stabilizes, you'll pay no attention to the floating sensation and get on with the work and daily tasks of living in space. After two days in space, I didn't even notice that I was living in free fall.

7. How does free fall affect your body?

After shedding what it thinks is excess fluid, which takes a couple of days, your body is quite happy in free fall with about a quart less blood volume than on Earth. At this new equilibrium, your legs appear skinny; astronauts call this "chicken legs syndrome." Sinuses stay congested for a week or so, perhaps causing a few headaches. Absent the usual compression from weight, your spine stretches an inch or more; this strains lower back muscles and can cause stiffness or an occasional ache.

The lack of weight on the skeleton causes your bones to start losing calcium, about 1 percent of your bone mass every month. Your heart and lungs would get lazy, too, but regular exercise prevents that. Finally, your immune system becomes less aggressive in responding to infection. Researchers are studying crewmembers staying on the International Space

Station for months at a time, hoping to find strategies that will keep astronauts healthy on expeditions to Mars. On these trips, astronauts will have to spend a total of a year or more in free fall.

8. How do you move around or anchor yourself in free fall?

It doesn't take much effort to move around inside a spacecraft when you feel weightless. On the space station, astronauts get proficient at moving across a module with just the press of a finger or a simple twist of the wrist. You don't have to think about how much force to push with—it's as instinctive as walking across a room.

A bigger challenge is how to stabilize yourself comfortably in front of a computer or an experiment, or when eating a meal in the galley. Two techniques used are tucking a foot under a handrail on the floor (or ceiling), or slipping your toes into a fabric loop anchored to a panel. On the treadmill, you wear a shoulder and waist harness attached to the machine with elastic tethers that pull you down onto the deck for running.

The author glides down the aisle of the Destiny lab on the ISS. Note the handrails on all four walls.

9. Can you experience something similar to free fall on Earth?

The most realistic free fall experience on Earth is found aboard NASA's McDonnell-Douglas C-9 research aircraft, called the "Weightless Wonder." Its famous predecessor was a Boeing KC-135, aptly nicknamed the "Vomit Comet" because of how some passengers reacted as the jet flew a succession of free fall maneuvers.

The C-9 first dives to pick up speed, then zooms into a steep climb that crests in a smooth arc "over the top." The pilots then push the plane into a dive. This climb-and-dive path is similar to that of a rollercoaster, or a cannonball arcing through the sky. This maneuver enables those in the plane's cabin, protected from air drag, to experience free fall for 20 to 25 seconds. The pilots then pull up into another steep climb, followed by another dive, repeating the maneuver as many as 40 times.

You would think that falling skydivers would also experience free fall, but air resistance, or windblast, slows their fall to a steady descent rate (called "terminal velocity"). Although skydiving gives a wonderful sense of freedom, it is not true free fall.

10. What fun things can astronauts try in free fall?

Tuck your knees to your chest and flip head over heels in an endless series of somersaults. Zoom through the modules of the International Space Station like Superman, challenging yourself to pass precisely through the center of the next hatchway. Have a friend park you in the center of an ISS module—out of reach of any handrails—and then try to propel your body back to a handhold. Perch upside-down on the ceiling like a bat while eating your lunch. Gobble a few floating bites of each meal right out of the air. Shoot malted milk balls through an improvised blow gun into a friend's open mouth. Carefully squeeze a glob of water the size of an orange from your drink bag, and scoot it around the cabin with puffs of your breath. When you retire for the night, make the wall or ceiling your personal sleeping quarters. Go to the bathroom upside down!

The author (center) experiences free fall on NASA's KC-135 aircraft. (NASA)

11. Is the air inside the ISS the same composition as air on Earth?

The ISS life support system mixes just the right amounts of oxygen (21 percent) and nitrogen (79 percent) to be almost the same composition as the air we breathe on Earth. The system leaves out trace Earth gases like argon and carbon dioxide.

Oxygen and nitrogen are stored in high-pressure gas tanks outside the *Quest* airlock, and in tanks delivered by Russian *Progress* cargo ships. These are tapped periodically to replenish the oxygen and nitrogen breathed in by the crew or lost through the airlock during spacewalks. Water vapor and carbon dioxide exhaled by the crew are removed by the life support system. The water is recycled into drinking water and oxygen, and the carbon dioxide is either dumped overboard or combined with humidity removed from the air to create more oxygen.

12. Where do you get the oxygen you need in space?

An astronaut needs 1.75 pounds (0.8 kilograms) of oxygen each day. On short missions of a few weeks, life support systems can mix oxygen and nitrogen from storage tanks to supply breathing air for the crew. In fact, we don't need nitrogen to breathe, but it is added to reduce the dangers of fire in a pure oxygen atmosphere.

On the International Space Station (ISS), those gases are stored in high-pressure tanks and fed into the atmosphere as needed. As a backup, the station stores Russian-supplied canisters of solid lithium perchlorate, which in a chemical reaction releases heat and gaseous oxygen.

To reduce the amount of breathing gas brought from Earth, the ISS life support system makes its own oxygen by pulling it out of the humidity in the air and from urine "contributed" by the crew. A separate

The Quest airlock at the ISS stores oxygen and nitrogen in the white high-pressure tanks with the horizontal handrails at left. (NASA)

system, called a Sabatier reactor, uses recycled water and the carbon dioxide exhaled by the astronauts to make more oxygen.

13. How do you heat and cool the inside of the International Space Station?

Space is a harsh environment, with sunlit temperatures rising to 250° F (121° C), and nighttime temperatures dropping to -200° F (-129° C). The outside of the ISS is covered by white or silvery coatings that reflect the Sun's rays to reduce the amount of solar heat that seeps inside. Insulation blankets wrapping the hull keep solar heat from penetrating and also reduce heat loss during the frigid night.

Inside the ISS, warmth from electronics and from crewmembers' bodies would quickly overheat equipment and make the station uninhabitable, so thermal control systems are a must. Chilled water circulates through equipment mountings and the ship's air conditioners to remove heat from cabin air, electronics, and lab experiments. The warmed water is pumped through a heat exchanger, which transfers the heat to ammonia-filled cooling lines leading to radiator panels that release it to empty space.

14. What is the temperature inside the International Space Station?

It's always shirt-sleeve weather on the ISS, about 72° F (22° C). Astronauts control the temperature and humidity inside using life support software on their laptops. The climate inside the station is also monitored constantly by Mission Control.

When I arrived on the station, I thought it was warmer and cozier than the shuttle. I was comfortable in a polo shirt and trousers. However, when I slept inside the station's new *Destiny* laboratory, the air conditioning was so efficient that I had to put on a sweater inside my sleeping bag to keep warm.

15. How can you make electricity in space?

A supply of electricity is vital for crew survival and to power research experiments in space. Batteries we bring from Earth can supply electricity, but they are heavy and only last a few days unless recharged. But solar energy is readily available in space, which is why the space station is equipped with large solar panels.

The surface area of the solar panels attached to the ISS is one acre (0.4 hectares). They provide from 75 to 90 kilowatts of electric power to run the station and to recharge batteries needed for the roughly 45 minutes of darkness during each orbit.

The *Gemini* and *Apollo* spacecraft and the shuttle produced power with fuel cells, which convert stored oxygen and hydrogen into electricity and water. Some planetary probes that are too far from the Sun to use solar energy generate electricity from the warmth of radioactive plutonium carried onboard. Nuclear reactors may generate electrical power for future spacecraft en route to Mars.

16. Is it noisy in space?

On the International Space Station, air circulation fans and fluid coolant pumps produce a constant level of background "white noise" despite sound insulation built into the equipment to reduce sound levels. Over time, this noise can damage an astronaut's hearing.

Crewmembers can wear earplugs or noise-canceling headphones for protection. Sleeping in their soundproof crew quarters while wearing earplugs gives astronaut ears a chance to rest and recover. Regular audiometer tests help detect and prevent any hearing loss.

Crewmembers regularly send sound level meter readings to Mission Control to keep track of noise levels inside the station. Any change in noise level might indicate a faulty fan or pump, or even a leak in the ship's hull.

17. Do astronauts feel crowded or claustrophobic in space?

There are 32,333 cubic feet (916 cubic meters) of laboratory and living space inside the International Space Station, equivalent to the space inside a Boeing 747 cabin. Astronauts have plenty of elbow room, and they can always look out through the cupola windows to a horizon a thousand miles away.

The *Soyuz* spacecraft has just 141 cubic feet (4 cubic meters) of habitable space, and you sit strapped in with your knees drawn well up toward your chest. The *Orion* and low Earth-orbit space taxis will be more spacious than *Soyuz*, but their crews will still be quite cozy aboard.

Our shuttle simulators and hundreds of hours in a snug Air Force B-52 cockpit prepared me well for the cramped atmosphere of a spaceship, and I never felt confined.

18. Do astronauts get homesick?

In space I was too busy to feel true homesickness, even as I realized the immense physical gulf between my family and me. Weekly radio or video talks with my wife and children helped a lot.

Astronauts on months-long expeditions certainly grapple with homesickness and family separation, just as our service members do when they are deployed overseas. E-mail and video messages are very important in combating feelings of isolation. But by far the best weapon against homesickness for ISS astronauts is their ability to phone Earth whenever they are off duty, catching up on events and goings-on back home.

19. Do astronauts get bored in space?

On my shuttle missions, I was too busy to ever even think about being bored. But astronauts on earlier space stations, and even some on the International Space Station, have become bored or depressed during their months-long stays in orbit. On future deep-space voyages, the best way to prevent space boredom is to provide astronauts with meaningful and engaging work to do, such as scientific research, astronomical observations, and analyzing samples collected from other planets.

20. Do astronauts get tired in space?

Astronauts work hard in space, both physically and mentally. My orbital workdays were 16 hours long, a period that included getting ready for work, meals, housekeeping, exercise, and winding down after work. My muscles got tired, especially after a spacewalk, but I think it was mental concentration that tired me most. To recover, I tried to get six to seven hours of sleep each night. My work in space was always exacting—I had never before been under such continuous pressure to perform perfectly. Although each of my missions was exhilarating, I did look forward to a vacation after landing.

21. How long are typical space missions?

An ISS crewmember spends about six months in space on a typical expedition. NASA astronaut Scott Kelly and Russian cosmonaut Mikhail Kornienko began a year-long stay at ISS in March 2015, investigating human health, endurance, and performance for eventual missions to Mars.

A typical shuttle mission lasted about 10 days to 2 weeks. The shortest shuttle mission was STS-2, which lasted 2 days, 6 hours, and 13 minutes. The longest was my third flight, *Columbia's* STS-80 mission—17 days, 15 hours, and 53 minutes.

22. Why is duct tape important in space?

Duct (or gray) tape is an indispensable invention that greatly eases the challenges of working and living in free fall. Its stickiness and strength make it a versatile helper in orbit.

Here are ten things you can do with duct tape:

- Seal bags of equipment for return to Earth.
- Reinforce and seal a plastic trash bag full of used meal packages.
- Tape down cables snaking around the modules.
- Repair broken equipment.
- Clean the lint from experiment air intake filters.
- Create labels to stick on storage bags or lockers.
- Hold down computer memory cards or USB memory sticks.
- Attach a duct tape tag, with reminder note, to critical switches or controls.

- Tape extra foothold loops to the floor.
- Stick loose food scraps on a loop of tape that can be rolled up and thrown away.

We usually launched with two or three large rolls stored in a locker. My crewmates joked that when you ran out of duct tape, it was time to come home. On ISS, crews also use the less-sticky Kapton tape, which doesn't leave a hard-to-clean adhesive residue on cabin surfaces.

23. How is the International Space Station an improvement over *Skylab, Salyut,* or *Mir?*

The International Space Station benefits from lessons learned from the *Skylab*, *Salyut*, and *Mir* space stations. *Salyut* could host two crewmembers; *Skylab* and *Mir*, three each, while ISS has plenty of room for six or more. It has eight living modules, two bathrooms, a mini-gym, two galleys, and a wrap-around bay window. The ISS's crew of six can also operate more sophisticated science experiments inside and outside than was possible on any of the previous stations.

The ISS should continue operating until at least 2024. Discoveries made on board are helping improve life back on Earth, and new technologies being tested there are helping pave the way for deep-space travel. For examples see NASA's Space Station Research and Technology website: nasa.gov/mission_pages/station/research/index.html

CHAPTER VI

Living in Space

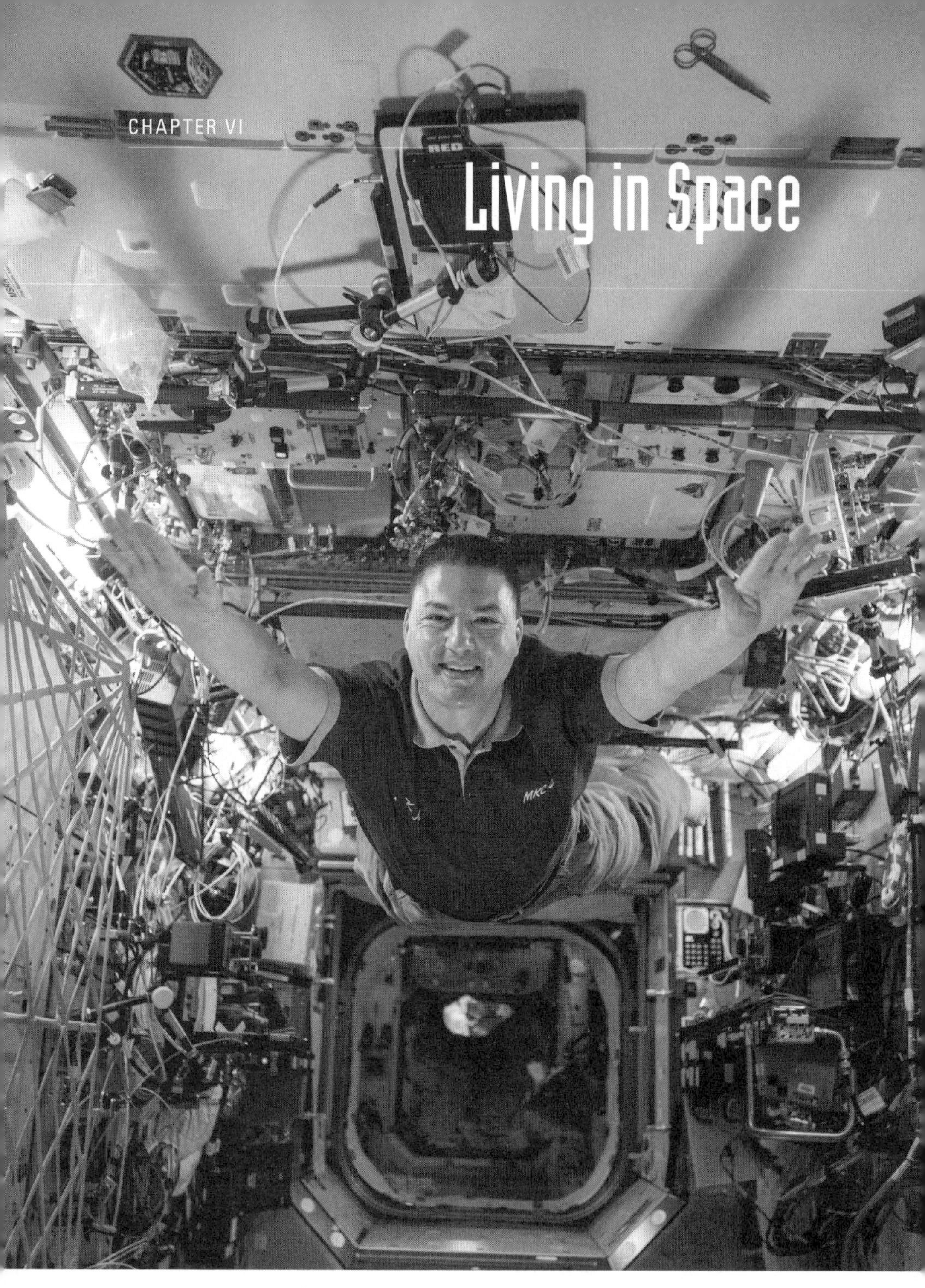

NASA astronaut Kjell Lindgren enjoys the freedom of free fall in the U.S. Destiny Lab aboard the ISS. (NASA)

1. Where are the kitchens on the International Space Station?

The kitchen on the ISS, as on a ship, is called a galley. That's where the crew prepares and eats its meals.

The Russian *Zvezda* module on the ISS has a galley with a water dispenser, food warmer, utensils, and clean-up supplies. A few yards down the central passageway is the U.S. *Unity* module, which has a galley and pantry stocked with breakfasts, desserts, snacks, pasta, meat and fish entrees, side dishes, vegetables, soups, and beverages.

NASA astronauts use a water dispenser and an electric food warmer to prepare their food around a modest table. The crew can chill their drink pouches in a small refrigerator. There's no need for chairs or knives: crewmembers float weightless around the table, and all they need are scissors and spoons to open and eat from meal packages. Velcro and bungee straps hold meal pouches on the table.

The *Orion* spacecraft will have a small galley for meal preparation on missions lasting several weeks. Low-Earth-orbit craft like *Soyuz*, *Crew Dragon* and the *CST-100 Starliner* reach the ISS within a day or two, so crews don't need a full galley—they just eat pre-packed meals until they dock at the station.

An ISS crew gathers for a meal in the galley of the ISS Unity module. (NASA)

2. How is food stored safely aboard the International Space Station?

On months-long trips aboard the ISS or into deep space, food must be stored safely so it doesn't spoil. Refrigeration and freezing can preserve foods but consume a lot of electricity that is better used to operate lab experiments or other vital spacecraft systems.

NASA uses several methods to keep food from spoiling but still preserve its appeal. Freeze-drying prevents bacterial spoilage by removing water from food. To prepare a freeze-dried dish, astronauts squirt water into its plastic pouch to rehydrate the food. In another process, called thermostabilization, meals are sealed in foil pouches and then cooked at high temperature and pressure to destroy bacteria. A third way to destroy bacteria is to expose packaged meals to a radiation source.

3. Do astronauts eat from a plate?

Crews on the International Space Station eat and drink right out of disposable plastic or foil containers—no washing dishes! When I fixed dinner in orbit, I occasionally used a metal tray with Velcro patches to hold down food packages. But the Velcro dots on the packages stick just as well to the Velcro squares on cabin surfaces or clothing. Magnetic scissors and spoons stick to metal strips on the galley table.

I typically ate from just a couple of opened packages at a time, being careful not to bump them and jar their contents loose. Opening more than a couple of items at a time is just asking for trouble—too many open packages is a recipe for a galley disaster. When you finish those first two, you open up the next pair. One of the simple pleasures of returning to Earth was enjoying an entire meal of favorite foods served appetizingly on a plate.

4. Is there a freezer or refrigerator on the ship?

Just a small fridge for chilling drinks, but crews occasionally receive ice cream sent up in medical research freezers onboard visiting cargo ships. The ice cream quickly disappears, and the freezers are then packed with blood, urine, and other biological samples for scientists to examine back on Earth.

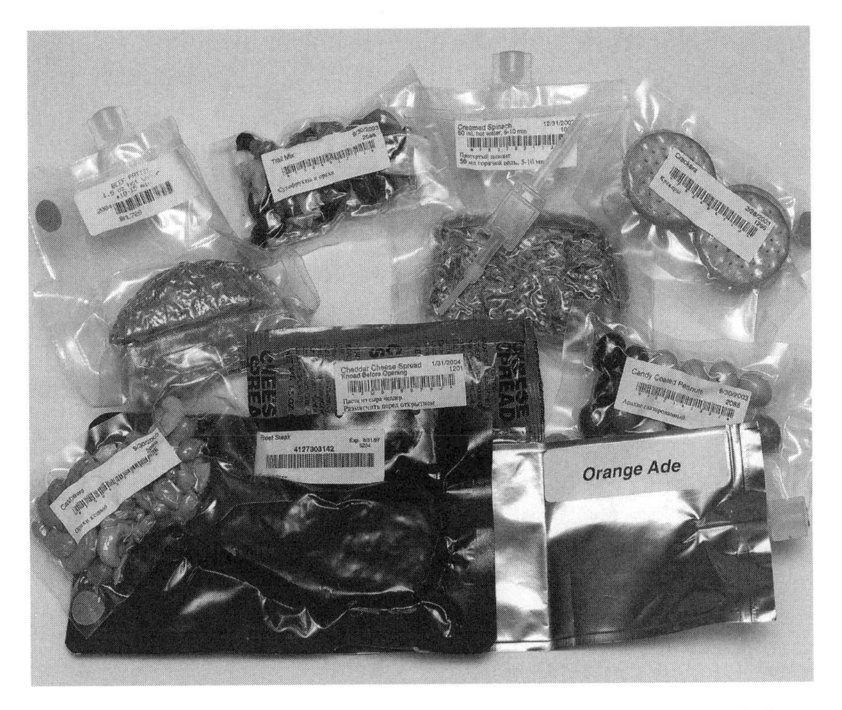

Space Station foods used for a typical Thanksgiving meal in orbit. From left to right: cranberry sauce, cornbread stuffing, smoked turkey, tea, and freeze-dried strawberries. All you'd need to enjoy them are scissors and a spoon. (NASA)

5. What nations supply food for the International Space Station?

NASA provides a full menu for its own astronauts and those of its European, Japanese, and Canadian partners. These partner astronauts also enjoy delicacies from their own countries launched via cargo ship.

The Russian space agency delivers food to its cosmonauts using *Progress* cargo ships and *Soyuz* spacecraft.

Crewmembers freely swap food items, especially when they get together for dinner. A varied menu adds interest to the meal, with the astronauts and cosmonauts sharing favorite foods and new tastes with their friends. NASA provides over 180 food and beverage items with Russia adding well over a hundred more, so ISS astronauts can easily avoid falling into a meal-time rut.

A meal of food packages laid out in the galley of the American Unity module on the ISS. (NASA)

6. How has space food improved over time?

Space food has definitely improved since the Apollo days of Tang and bite-sized bacon cubes. And since the shuttle era, improved food preservation techniques have greatly increased the variety of textures and flavors available. Astronauts today choose their meals from a much more varied menu.

About five months before launch, astronauts sample the complete menu and put together a series of daily menus that repeats about every ten days. At any time, you can choose a snack or vary your daily menu with a different item from the galley pantry.

Dropper bottles of salt water and pepper grains in oil are handy to season foods, and you can spice things up with condiments such as individual packets of ketchup, mustard, and Tabasco, taco, picante, and steak sauces.

Periodic shipments of fresh food arriving via cargo ship enhance the menu every two or three months. Sadly, you still can't get a good pizza or burger "your way" in space.

7. What kind of foods did you enjoy in space?

My favorites included breakfast burritos, Kona coffee, chocolate brownies, smoked almonds, oatmeal with brown sugar, grilled chicken wrapped in a tortilla, freeze-dried strawberries, chocolate pudding, lasagna, spaghetti in meat sauce, freeze-dried asparagus, spicy chicken in sauce, macaroni and cheese, grilled steak, and sliced barbecue beef.

I always liken my space food experience to eating meals on a camping trip. Food might be dehydrated or packed in foil pouches, making it easier to prepare and serve than fresh-cooked meals, but it is less satisfying than home cooking. I missed the textures and flavors of fresh vegetables and fruits, salads, and the different aromas of several foods together on a plate.

The author chows down on a "chicken flying saucer sandwich," an irradiated chicken breast, glued between warm tortillas with picante sauce. (NASA)

8. What favorite foods can you bring from home?

Astronauts on the International Space Station can request favorite snacks be shipped up to them in bonus food containers, which arrive with other fresh foods aboard cargo ships or space taxis like *Soyuz*, *Crew Dragon*, and the *CST-100 Starliner*.

Each astronaut on my crews could request a favorite food item, and NASA's dieticians would add it to the shuttle's fresh food locker. One astronaut chose chocolate chip cookies, which were tasty but generated a lot of crumbs. Another brought a plastic bag full of bite-sized Swiss chocolates. The snacks I chose were the ones I carried in my lunch box when I was a boy—Tastykake cupcakes and snack cakes called Krimpets. These were safe at room temperature and moist enough to minimize crumbs. I ate one every day and traded extras for other menu items.

On my shuttle mission to the ISS, my crew also requested Maryland crab soup, freeze-dried in individual packages. Its spicy, savory flavor proved so popular that we left some with the station crew when we returned to Earth.

9. Do astronauts gain weight in space?

I sure didn't. The freeze-dried and thermostabilized meals on the ISS, together with ready-to-eat foods like nuts and dried fruit, provide a balanced diet of 1900 to 3200 calories per day, depending on astronaut size and gender.

That sounds like a lot of calories, but astronauts on the ISS exercise strenuously every day, and that routine combined with the flavors and textures of space food means that some weight loss is likely. ISS crewmembers living in orbit for six months lose an average of from 4 to 9 pounds (2 to 4 kg) for men, and 3 to 7 pounds (1.3 to 3.2 kg) for women. Shorter missions don't result in much weight change. I usually came home within a pound or two of my launch weight.

10. How do astronauts weigh themselves in free fall?

In free fall an astronaut's weight is essentially zero, all the time. But astronauts still have mass, and by measuring their mass in space, they can determine what they would weigh on the ground. Flight surgeons ask for regular mass measurements to ensure astronauts are eating properly and staying healthy during their six-month expeditions on the International Space Station.

One mass measurement device aboard is the Space Linear Acceleration Measurement Device (SLAMMD). It uses Newton's second law of motion (F=ma) to determine an astronaut's mass. SLAMMD uses two springs to apply a known force to the astronaut, and then measures the resulting acceleration to figure out mass. The readings are accurate to within half a pound. A Russian device also measures mass while an astronaut rocks back and forth on a spring-loaded, pogo-stick-like platform.

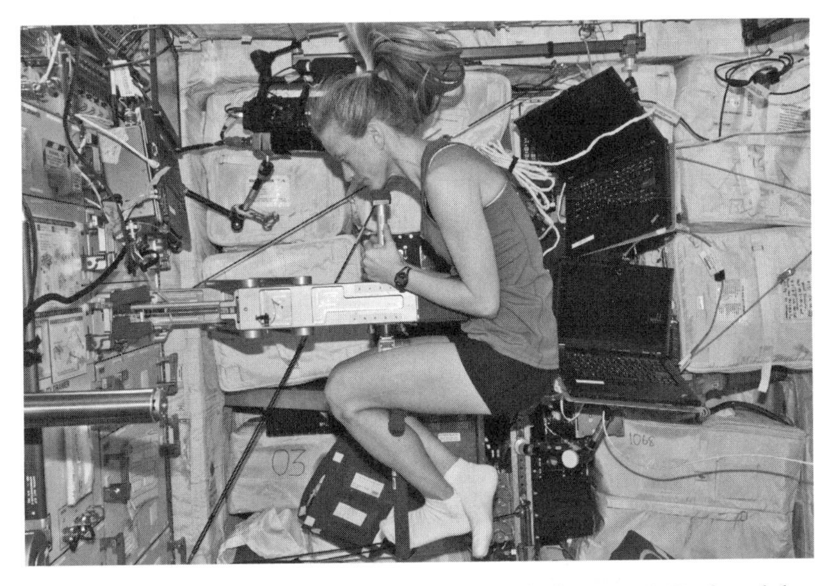

Astronaut Karen Nyberg checks her body mass with the SLAMMD aboard the ISS. (NASA)

11. How is water supplied and recycled on the International Space Station?

Each astronaut needs about 7.7 pounds (3.5 kilograms) of water each day (nearly a gallon of water a day). That's over 2,800 pounds per year. Fresh water is sent to the ISS in tanks aboard supply spacecraft such as *Progress*, *Dragon*, *HTV*, and *Cygnus*.

To reduce the cost of shipping water from Earth, the ISS life support systems recycle waste water. Water vapor exhaled by the crew is captured and fed into a water recovery system, and then stored as drinking water or broken down to make breathable oxygen. The crew's urine is reclaimed from the toilet system, fed into the urine processor where it is purified and then added to the drinking water supply. This system recycles about 70 percent of the urine produced by the six-person crew.

Recycling water saves tens of millions of dollars in reduced launch costs every year. To keep the spacecraft weight (and thus cost) down on a Mars expedition, water recycling systems will have to be even more efficient.

12. How do you go to the bathroom in space?

I thought you'd never ask! In our Earthly bathrooms, gravity makes everything go where it's supposed to go. To design a toilet that works in the weightless conditions on the International Space Station, engineers used moving air as a substitute.

Electric fans pull air through the toilet system to draw liquid and solid waste away from your body. Astronauts flip a switch to turn on the toilet fan. They then attach a funnel to the top of a vacuum hose and hold it close to their bodies. Air rushing into the hose moves urine into the funnel and carries it along to a storage tank.

Hooking toes under a floor rail, astronauts position themselves just above the toilet seat. Air flows under the seat and draws solid waste through an opening and down into a plastic pouch. After each use, waste, toilet tissue, and sanitary gloves are sealed into the pouch and pushed down into a canister under the seat. Finally, the user cleans the seat and canister mouth and installs a new pouch, readying the toilet for the next visitor. Sterilizing hand wipes go into a separate trash bag.

13. What happens to the waste?

Urine is stored temporarily in a tank and then piped into the water recovery system, where the urine is recycled into clean drinking water.

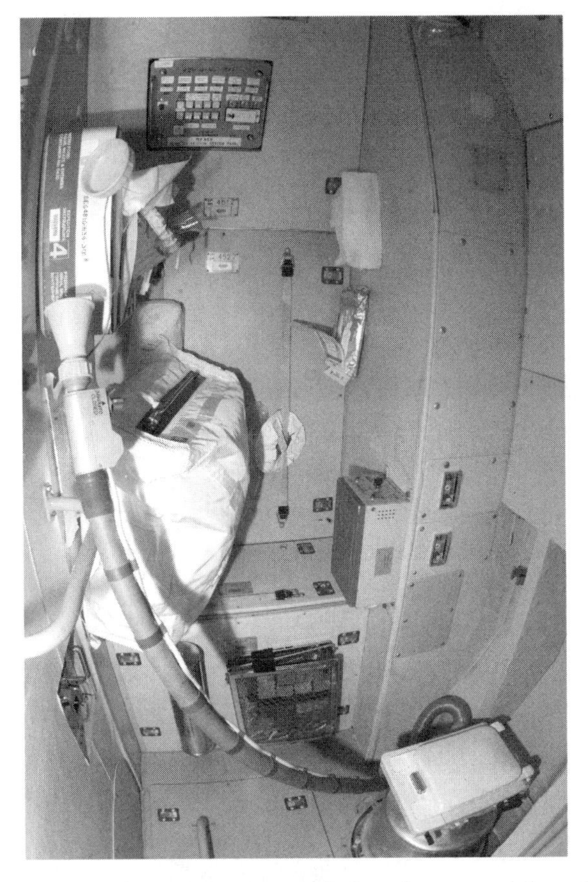

Seen here is an ISS toilet compartment with funnel at upper left, and the seat with closed lid at right. (NASA).

The life support system turns some of that clean water into breathable oxygen.

Astronauts remove the solid waste canister from under the seat when full and store it with other trash in empty cargo ships. When those ships leave the International Space Station, they fire their retrorockets to reenter the atmosphere and burn up, incinerating the waste. Next time you see a shooting star streaking across the night sky, remember—it might be a gift from the astronauts!

14. How do astronauts exercise in space?

For aerobic exercise, astronauts run on a treadmill and ride an exercise bike. But to keep them from floating off these machines while weightless, they have to wear special equipment. An elastic harness attached to the waist and shoulders holds the runner on the treadmill. On an exercise bike, astronauts wear a loose waist belt looped around a stabilizer strut and snap their cycling shoes into the pedals. I've pedaled the shuttle's exercise bike all the way 'round the world'—in just 90 minutes!

Crewmembers on the International Space Station put in regular strength training sessions on the Advanced Resistive Exercise Device. This weight-lifting machine uses the difference between the cabin's air pressure and the natural vacuum of space to place a resisting force on a pair of pistons, enabling astronauts to grip a bar and perform familiar curls, squats, and lifts. Bungee cords, elastic stretch bands, and spring-loaded grippers round out the exercise options.

NASA astronaut Dan Burbank lifts "weights" on the Advanced Resistive Exercise Device at the ISS. (NASA)

After exercising 90 minutes a day for six months or more, most astronauts return with excellent muscle tone, and have greatly slowed their rate of bone loss. This is all good news for long Mars missions.

15. What happens when you sweat in space?

On Earth, warm air rises because it is less dense than the cooler air surrounding it, which slides in beneath to replace it. But in free fall, there are no density differences, so air warmed by your body does not rise. When you exercise, your body heat just warms the air around you, wrapping you in a blanket of hot, humid air. Because of this layer of insulating, humid air next to your skin, perspiration doesn't evaporate easily. Sweat actually puddles on your skin. Astronauts deal with this extra perspiration by toweling off frequently, and directing a vent hose or a fan-driven stream of cool air toward themselves while exercising.

Astronaut Sunita Williams runs on the ISS treadmill during Expedition 32. (NASA)

16. How do you keep clean in space?

You might think that it's hard to get dirty in a spaceship. But much of my work on the International Space Station was physically demanding. I hauled cargo from shuttle *Atlantis* to the station (it had mass, even if it *was* weightless), installed safety equipment, wiring, and radio gear in the new *Destiny* lab, and worked outside on three spacewalks. ISS crews work just as hard, and the astronauts exercise for at least 90 minutes every day. If they don't want their colleagues to flee, they have to bathe.

On *Skylab* in 1973–1974, we learned that it took astronauts too much time to dry off and vacuum up the watery mess from a space shower. So on the ISS, astronauts squirt hot water and rinseless soap onto a washcloth and wipe down from head to toe.

Each astronaut has a toiletry kit, complete with Velcro patches inside to hold toothpaste, toothbrush, dental floss, clippers, lip balm, shampoo, and a razor. Battery-powered razors capture whiskers and hair in the shaving head. Investing 15 minutes once a day in personal cleanup ensures you will have friends who don't mind staying close to you in space.

Astronaut Karen Nyberg washes her hair on the ISS. (NASA)

17. How do astronauts wash and trim their hair?

To wash their hair, astronauts first squirt warm water from a drink bag onto their scalp. They then add rinseless shampoo (the same kind used in hospitals), and work it into their hair with fingers or a comb. After drying their hair with a towel, they comb it into shape. On the International Space Station, astronauts take turns trimming each other's hair with scissors, using a vacuum cleaner to catch the clippings.

18. How do astronauts brush their teeth?

Astronauts floss their teeth like they do on Earth, but brushing is harder—there's no sink to spit toothpaste into! Astronauts spit toothpaste into a tissue or towel, and then rinse their mouths with clean water. To save water, astronauts on the International Space Station have begun using edible toothpaste, swallowing it after brushing.

19. Where do you sleep in space?

On the International Space Station there are individual bunk compartments for each crewmember. In the *Harmony* module, there are four crew compartments built into the walls, ceiling, and floor; another two are in the *Zvezda* service module.

The compartments each have room for a sleeping bag, clothing storage, a laptop desk, an intercom, and wall space for photos and mementos. Each has an accordion-style door that closes for privacy and soundproofing; the Russian crew quarters have windows with a view of Earth.

On spacecraft headed to or from the ISS, astronauts clip their sleeping bags to the wall, floor, or ceiling. They wear an eyeshade to keep out stray sunlight, and snooze away in their temporary bedroom. Closing my eyes, I always felt as if I were sinking into the softest, deepest feather mattress that ever existed, falling asleep within minutes.

20. What privacy do you have on the International Space Station?

Living six months or more on the station with five other space travelers might sound like a crowded proposition, but the ISS has a

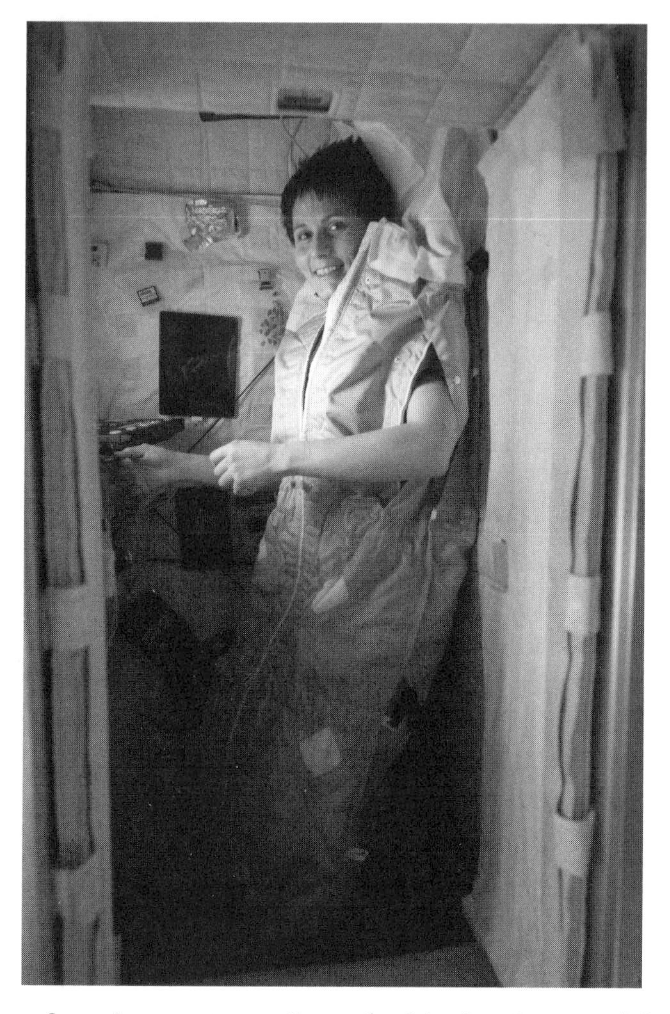

European Space Agency astronaut Samantha Cristoforetti rests inside her personal crew quarters on the ISS. (NASA)

roomy feel. You can work for hours in one location and not see anyone else because your crewmates are working on other tasks elsewhere.

You can always find a private moment in the station's airlocks, storage modules, or the cupola.

Astronauts change clothes in their sleeping compartments, and exercise and bathe in the *Tranquility* or *Zvezda* modules. Each crewmember can retreat to his or her personal compartment, and everyone does their best to respect personal privacy.

21. Do astronauts do house cleaning in space?

Homes in space need cleaning and maintenance, just like those on Earth. Astronauts spend a part of each day doing chores, such as wiping down the kitchen, vacuuming lint from the air filters, emptying trash containers, and cleaning the bathroom.

Once, while cleaning some air filters behind a panel on the orbiter flight deck, I found a drifting hairbrush and a "Go Air Force!" bumper sticker. These items, lost by the previous crew, had drifted into the tiny space and lodged there.

By performing routine maintenance and fixing broken equipment with spare parts and help from Mission Control, astronauts keep things humming high overhead.

22. What happens to your trash?

Wastepaper, trash, tissues, used duct tape, and packing material are called dry trash. It's discarded into familiar plastic trash bags that are tied or taped closed. Wet trash is anything that will soon smell bad: used food packages, wet paper towels, and hand wipes. These are placed in plastic bags, compressed by hand, and sealed with duct tape.

Along with canisters of the crew's solid bathroom waste, the trash bags are stowed in empty cargo ships like the *Progress*, *Cygnus* and *HTV*. Once full, these spacecraft undock and are guided into the Earth's atmosphere, where they burn up. You could say the garbage is recycled down to its very molecules during reentry.

For deep-space missions, NASA is exploring ways to recycle trash for use as propellants, breathable oxygen, and radiation shielding.

23. Do spaceships have a distinctive smell?

Apollo astronauts reported that Moon dust tracked into the lunar module gave off the faint, acrid smell of gunpowder. After a spacewalk, I noticed that our suits had the pungent, burnt electrical odor of ozone.

While outside, stray oxygen atoms had stuck to our spacesuits. Those atoms combined with oxygen molecules in the airlock to form ozone, resulting in that pungent odor.

The space station life support system has an odor control system called the Trace Contaminant Control Subassembly. It consists of an activated charcoal bed, a catalytic oxidizer assembly, a lithium hydroxide absorbent bed, a fan, and a flow meter.

The charcoal bed removes ammonia and most other odors, but gases like methane are trapped in the catalytic oxidizer and destroyed by heating them to 750° F (400° C). The compounds and acidic gases produced during this heating process are absorbed by the lithium hydroxide bed. The system handles nearly all of the food, body, trash, and waste odors produced by the crew. When I visited the International Space Station, the outpost smelled clean and fresh, and its air filtration systems continue to perform well today.

24. How do astronauts do their laundry?

They don't. Doing laundry in free fall would require a specially designed washing machine that would use a lot of water and power, draining the station's limited reserves.

Crewmembers on the International Space Station change their underwear and socks once every two days and their shirts and pants (or shorts) once a month. They change their sleepwear—a T-shirt and shorts—once a week and then wear them during exercise for another week. The soiled clothing simply goes into the trash, which is dropped into Earth's atmosphere on a cargo ship to burn up.

By contrast, shuttle clothing was luxurious. I had a fresh shirt, socks, and underwear every day, and much of that clothing was washed back on Earth and used on later missions. Space clothing designers are experimenting with clothing items that have been impregnated with bacterial inhibitors to prevent odors, so clothes can be wearable and bearable for longer periods.

25. What do astronauts do for fun in space?

Just living and working in space was fun—the view, the friendships, the fascinating work, the freedom of living weightless in free fall. I've never worked as hard as I did when I lived in orbit, but I went around smiling all day long.

When I did have spare time, I stayed glued to a window, camera in hand, snapping away at Earth landscapes so I could share them with geologists, Earth scientists, and family back home. When not looking out the windows, my crew admittedly had a blast zooming down the aisles of the space station and spinning a few dozen somersaults. On the International Space Station, watching recorded movies on the big-screen projection TV in the *Destiny* lab is always popular.

26. Do you have spare time in space, and how do you use it?

Astronauts must pace themselves on their long, six-month stays aboard the International Space Station. With workdays lasting 10 hours or more, taking some time off is important. Astronauts work a half-day on Saturdays, and Sundays are left to the crew for relaxation, various chores, and unstructured time. Most crews do put in some science time on Sundays, too, because they find the research work interesting and fun.

Each night after dinner, crewmembers usually have 90 minutes or so of free time before going to sleep. They can spend this time taking in the magnificent view from the ISS cupola, or pursuing hobbies like reading, listening to music, sketching, sewing, playing a musical instrument, or making contact with amateur radio operators on Earth..

27. Are there books on the space station?

Yes. Astronauts can read e-versions of their favorite titles on their personal tablets or laptops. On my last flight, I took along my copy of *2001: A Space Odyssey,* the classic sci-fi novel. The 1968 movie based on this book helped inspire me to become an astronaut. Early space stations like *Mir* actually had libraries with dozens of paperback books.

NASA astronaut Tracy Caldwell Dyson takes in the view from the ISS cupola. (NASA)

The author with 2001: A Space Odyssey aboard Atlantis during shuttle mission STS-98. (NASA)

28. Do astronauts play music in space?

Music is a bit of your life on Earth that you can enjoy while working or relaxing in space. Astronauts on the International Space Station listen to their digital music collections on iPods or laptops. They also enjoy music sent up to them via Mission Control from friends and family back home. While gazing down at Earth before sleep, I often listened to relaxing music, and I amped up my workouts with rock and dance tunes. A few astronaut-musicians have taken their instruments—guitar, flute, or keyboard—along on their expeditions as a way to unwind, be creative, and share music with their crewmates.

29. What did you miss while you were in space?

I missed the familiar feel of a breeze on my cheek, and the rich smell of mown grass or a summer forest on a camping trip. I missed favorite foods, like pizza, steamed crabs, and fresh fruits and vegetables, along with the sight and aroma of several delicious food items all sharing the same plate in front of me. During intense workdays on the shuttle, I missed reading for pleasure. I missed my wife and children. I was able to endure all these postponed pleasures because I knew our science and exploration work in orbit was important, and because their absence in space made my return to Earth even more memorable.

30. Can you watch TV in space?

Astronauts can watch DVDs or digital movies, sent up to their computers via Mission Control, on their laptops or project them onto a 65-inch screen. Crews have regular movie nights, gathered around the big screen with snacks and drinks. Sporting events can be beamed up, too, and recorded onboard for watching during free time. On weekends, though, the crew might have enough free time to watch some major events live, like World Cup soccer, the Super Bowl, or Olympic contests.

31. Will my cell phone work in space? If not, how do you call home?

Cell phones do not have the power to transmit to or receive signals

The 65-inch (1.6 meter) projection screen in the Destiny lab aboard the ISS. (NASA)

from cellular towers on the ground, 240 miles (389 kilometers) below. Instead, radio and video communications between the ISS and Mission Control go through geosynchronous communication satellites orbiting overhead at an altitude of 22,236 miles (36,022 kilometers). The ISS is out of range of those satellites for about 10 minutes of each 90-minute orbit, in zones where Earth blocks the view of the satellites.

Through this satellite link, ISS astronauts can reach Earth using laptop software with voice-over-internet-protocol (VOIP) programs similar to Skype, enabling them to dial their families, friends, and NASA colleagues whenever they would like. I used this system to call my family from shuttle *Atlantis*. The connection was clear as a bell.

Astronauts on the space station today can also call Mission Control on the radio, send emails, update their Facebook pages, and launch Tweets back to Earth.

Marsha Ivins speaks to family using the laptop-to-Earth radiotelephone. (NASA)

32. What were your favorite things about being in space?

The view of Earth was ever-changing, ever-beautiful, and ever-inspiring. I enjoyed the freedom of living and working in free fall. Despite its complications and annoyances, it was hard not to find it magical. I found it intensely satisfying to share my space experience with close friends, and serve as part of a team—in orbit and on the ground—working to achieve great things together.

33. How do astronauts celebrate holidays and birthdays while in space?

On the space station, astronauts enjoy celebrating individual birthdays as well as their international crewmates' national festivals and holidays. Crewmembers display their traditional holiday decorations sent up via cargo ships. For big holidays like Christmas, they get a day off work and call their families via video conference.

The crew will typically gather for a special dinner, play music of the season, and share delicacies or favorite holiday dishes sent up by supply ships. A U.S. Thanksgiving meal may include irradiated smoked turkey, thermostabilized candied yams, Russian mashed potatoes with onions, freeze-dried cornbread stuffing, freeze-dried green beans with mushrooms, and thermostabilized cherry-blueberry cobbler.

Gifts and special foods arrive in the equivalent of space goodie bags—bonus food containers in cargo shipments. These might contain videos, games, notes, and prized snacks like chocolates, candy canes or other holiday treats.

Birthdays call for a special dinner, small gifts or favorite foods, and greeting cards or video messages sent from home. As a bonus, Mission Control has been known to sing "Happy Birthday" to the lucky astronaut.

Expedition 42's Samantha Cristoforetti (from Italy) decorates for Christmas in the Destiny lab on the ISS, December 2014. (NASA)

34. How easy is it to write or type in space?

On the space station, astronauts have adjustable laptop desks in the labs and in their crew quarters, making typing easy. On the shuttle, I found it hard to stabilize my body near the laptop or to rest my wrists on

The author enters Earth observation data on a laptop aboard Endeavour during shuttle mission STS-59. (NASA)

the keyboard, so typing more than a sentence or two was awkward. With a floating hand, pen, and notepad, my handwriting was sloppy, too. For any detailed note taking, I would pull out my microcassette recorder and just dictate comments, then transcribe them after landing.

35. What will space tourists experience in orbit?

Orbital tourists will spend their days off the planet on visits to orbiting commercial space stations. They will enjoy tremendous views of Earth and the novel experience of getting used to living in free fall. They'll choose their own menus and enjoy eating a meal on the wall or ceiling. A space station will give them room to practice free-fall gymnastics. They'll sleep in individual cubicles or take a chance at sleeping adrift in their sleeping bags. The biggest difference from my own flights is that space tourists won't have to work 16 hours a day. But if anything like mine, their launch, orbit, and reentry experiences will be truly unforgettable.

CHAPTER VII

Working in Space

Astronaut-geologist Harrison "Jack" Schmitt collects lunar samples during the 1972 Apollo 17 mission, the last human expedition to visit the Moon. (NASA)

1. What sort of work do astronauts do in space?

An astronaut is very much a jack-of-all-trades, trained to do a little bit of everything well. Astronauts assigned to the International Space Station pilot their ships to and from the outpost and, once on board, conduct medical experiments, maintain and repair their station, do laboratory research, install new equipment, and perform spacewalks.

In orbit, I was a radar imaging system operator, shuttle flight engineer, spacewalker, ISS construction worker, robot arm operator, satellite operator, lab technician, computer network manager, space rendezvous specialist, docking system specialist, cargo handler, Earth observer and photographer, photo and video manager, IMAX camera operator, cook, and custodian.

The author installs IMAX camera lighting equipment in the Zarya module at ISS. (NASA)

2. What is a routine day like on the International Space Station?

A typical day aboard the station looks like this:

06:00 AM: Crewmembers wake up. Wash up, eat breakfast, and read news and messages sent up from Mission Control overnight.

07:30 AM: Morning Planning Conference. Astronauts and cosmonauts sync up with Mission Control centers on Earth before beginning the planned activities for the day.

07:55 AM: Prepare for work, go over procedures, and set up any equipment needed for the day's activities.

08:15 AM: Crew starts work. Activities include conducting science experiments, maintaining and fixing equipment and systems on board, preparing for visits from other spacecraft, storing cargo, station environment sampling (sound levels, bacteria on surfaces and in the water), and conducting interviews with the media about station events. Crewmembers also exercise for about 2 hours and take medical measurements to record and assess their fitness over the course of the mission.

01:00 PM: Lunch

02:00 PM: Crew gets back to work.

06:15 PM: Prepare for the next workday. Review procedures and the schedule for the next day.

07:05 PM: Evening Planning Conference. Discuss the day's work with Mission Control, and make changes to tomorrow's plan, if needed.

07:30 PM: Eat dinner, relax, email, organize images to send back to Earth, call family, look at Earth!

09:30 PM: Crew goes to sleep for eight and a half hours.

3. How do astronauts know what work should get done each day?

Each day, Mission Control sends the crew on the ISS a flight plan, which shows each crewmember's schedule. The flight plan, radioed up to the crew's laptops, is quite specific. It details the activity start time through each crewmember's 15-minute block of time during the workday.

Critical tasks take priority, like changing the orientation of the station, docking an arriving supply ship, or starting a spacewalk. Because each astronaut spends about six months on the ISS, lower priority tasks that aren't completed can be rescheduled for the next day or week. If an astronaut finishes early, he or she can choose from a "job jar" of tasks that must be done eventually, but are not time-critical.

Astronaut Sunita Williams checks her ISS flight plan in the Destiny lab. (NASA)

4. Do astronauts help decide on the flight plan for their work?

Yes. The crew holds a conference with Mission Control each morning and evening to discuss the day's work and what's being planned for the next day. Flight controllers review the next day's priorities, and the astronauts make recommendations on how to get those tasks done more efficiently. Sometimes a flight plan changes during the workday to adjust to changing circumstances aboard the station.

These regular conversations keep everyone in sync, and help minimize misunderstandings. Good communication enhances the teamwork between the flight controllers and the astronauts.

5. How do astronauts keep track of time in space?

Telling time is central to an astronaut's life and work in space. A successful mission depends on accomplishing critical space tasks and research experiments on schedule. Because astronauts experience 16 sunrises and sunsets every day while in low Earth orbit, the Sun's position in the sky isn't useful for telling time. On deep-space missions, the Sun's very slow motion against the background stars won't help with timekeeping, either.

Instead, daily schedules in space are based on Coordinated Universal Time, or UTC, and astronauts set their watches and computers to this time.

We kept time differently during my shuttle flights. We tracked events in "mission elapsed time"—the days, hours, and minutes since liftoff, when the mission clock started. I set my digital watch display to the mission elapsed time, and used the hands on my watch face to show what time it was at home, back in Houston, where my family and Mission Control were located.

6. Can astronauts always talk to Mission Control in orbit?

Astronauts on the International Space Station talk to Mission Control through NASA communications satellites in geosynchronous orbit around Earth. These Tracking and Data Relay satellites are high enough to provide near-continuous coverage over most of the station's orbit.

Because the ISS must share these satellites with other users, like the *Hubble Space Telescope,* the station cannot always communicate through them. Laptop displays at the ISS show the crew when a radio or video link to Mission Control is available. Outages last for just five or ten minutes out of a 90-minute orbit.

As we head out into deep space, future crews will face delays in communication caused by distance. For example, radio waves traveling at the speed of light take nearly 1.5 seconds to reach the Moon. A one-way signal to Mars would take as long as 20 minutes, making normal conversation impossible. An astronaut on Mars would have to wait as long as 40 minutes to get an answer to a question!

7. Who is the boss on the spaceship, and who is responsible for accomplishing the mission?

Good leadership in space and at Mission Control is essential for a successful mission. At Mission Control, the flight director is responsible for the overall safety and success of the mission. The flight control team, working with the crew before launch, creates a safe and practical plan for accomplishing the mission. The spacecraft commander makes sure that the plan is safely and properly executed. Because decisions in space may be time-critical and carry life-or-death consequences, the commander has ultimate responsibility for the immediate safety of his or her crew.

8. Do astronauts ever have arguments or conflicts?

Astronauts naturally have occasional disagreements, even within a close-knit space crew. Getting along is particularly important for crews on long, deep-space missions. You don't want to encounter serious conflicts between crewmembers on a long voyage in a confined space, with no possibility for relief.

To minimize such conflicts before launch, instructors and managers will observe a long-duration crew closely during training to see how well the crewmembers get along. If serious personality conflicts arise, the Chief Astronaut can substitute a backup crewmember.

Deep-space crews will get along best if crewmembers have a few key traits. They should be comfortable both with working alone on projects and socializing together—no introverts or extroverts. Astronauts should recognize that they will have differences with others, but be willing to work them out. A sense of humor will be invaluable on a long expedition. I believe a successful crew should also include a decisive commander, one who invites opinions but doesn't hesitate to make the final call.

9. Are mistakes made while working in space?

No human being is perfect, so mistakes will happen. Fortunately, Mission Control is also run by humans, and they understand that everyone makes mistakes. When you make a mistake, you tell your crew and the ground about it; they help you recover by recommending what steps you should follow to correct the error. By working together and maintaining open communication, we keep mistakes and misunderstandings to a minimum. Don't worry—help is as close as a crewmate or a radio call to Mission Control.

10. Can astronauts repair things that break on their spacecraft?

Yes, we can make repairs on the International Space Station. The outpost is equipped with a wide range of tools and spare parts, ranging from screwdrivers and wrenches to complete ISS subsystems, like coolant pumps and solar array voltage regulators.

During our mission to the station, spacewalking partner Bob Curbeam and I were able to fix a nasty ammonia leak in a cooling system hose that supplied the *Destiny* laboratory. Bob acted quickly to close a valve that eliminated the leak. Working together, we were able to reinstall the hose and eliminate the problem.

A famous station repair happened in 2007, when two astronauts successfully fixed a torn solar array in a 7-hour, 19-minute spacewalk.

Astronaut Scott Parazynski was a member of the team that fixed a torn solar array panel at the International Space Station. (NASA)

They used spare wire, tape, and aluminum strips to create a set of reinforcing links that still hold one of the four solar array wings together.

11. What's in the toolkit on the International Space Station?

The station has a very well-organized and astonishingly comprehensive toolkit, one that most home handymen can only dream of. Among the items are: vise grips, pliers galore, socket sets (English and metric), nut drivers, an inspection mirror, a tape measure, tin snips, cordless drills, drill bits, a dead blow hammer (one that doesn't recoil in free fall), a crowbar, a fiber optic boroscope, torque tip drivers, screwdrivers, box-end wrenches, a wire bone saw, a hack saw, tweezers, wire strippers, cable cutters, clamps, long-handled forceps, metal files, chisels, electrical grounding straps, a monkey wrench, a flashlight, a headlamp, safety goggles, electrical tape, and a multimeter.

There's no bubble level—remember, this is free fall, and there is no up or down or level. If more tools are needed, ISS astronauts have already tested making new ones with a 3-D printer aboard the station. Oh—duct tape is used so often that it gets its own storage locker.

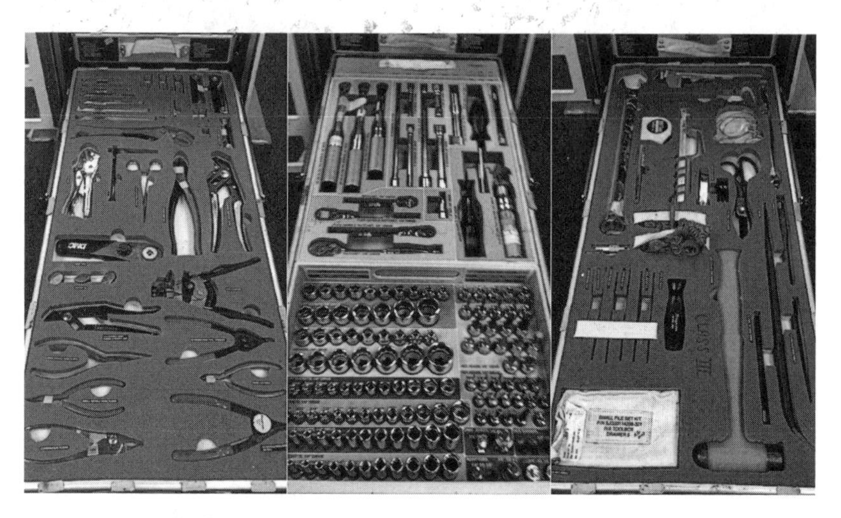

Several drawers of the toolkit aboard the ISS, seen in a NASA ground training mockup of the station. (NASA)

12. What is the most dangerous job on a space station?

What I consider the most rewarding task I did in space—performing a spacewalk—is probably one of the riskiest. On a spacewalk outside the International Space Station to install the U.S. *Destiny* laboratory, I worked in my spacesuit for anywhere from 8 to 10 hours. Alongside my spacewalking partner, Bob Curbeam, we worked in a deadly vacuum, exposed to the fierce heat of the Sun and the bone-chilling cold on Earth's night side. If my suit had failed, or if I had incorrectly operated its systems, I would have been dead in seconds.

Working in a stiff spacesuit, moving it around the station and handling tools with clumsy gloves for hours on end is a tough physical challenge. The mental pressure is just as severe. Making a mistake could damage equipment on the station and cost millions of dollars to repair, or impair its scientific mission. I had to concentrate hard for hours on end, trying to avoid making a mental error. Because of the exhilaration of working outside, along with the physical risks and the intensity of the work, spacewalking is both the neatest thing and the most dangerous thing an astronaut can do.

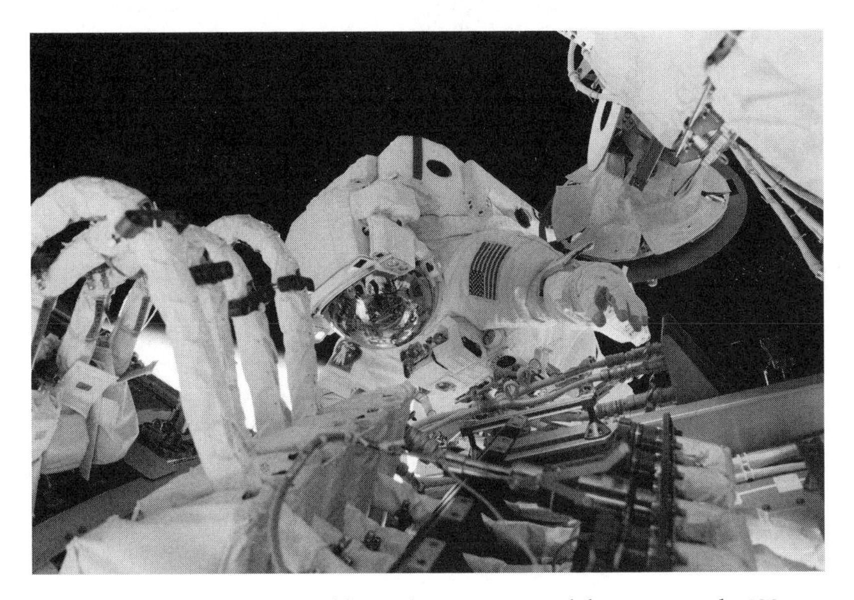

The author connects power cables to the new Destiny laboratory on the ISS. (NASA)

13. After a mission, how long before you can fly in space again?

NASA astronauts wait an average of five years between expeditions to the station. Astronauts returning from space are asked to contribute their experience and skills to help plan future missions, train other crewmembers, and help develop new spacecraft. They also spend another two-and-a-half years in training before heading for the International Space Station again.

14. Where did you go in space?

I often get asked if I went to the Moon, but I became an astronaut long after those Apollo expeditions to its surface. I traveled in space near our planet in low Earth orbit, a region stretching between 99 miles (160 kilometers) and 1,200 miles (2,000 kilometers) above the surface of the Earth.

NASA astronaut Randy Bresnik prepares to enter a mockup of the CST-100 Starliner spacecraft. (NASA)

During my visit to the International Space Station, it was at an altitude of 220 miles (356 kilometers). The Apollo explorers went a thousand times farther from Earth. With support from the President and Congress, the Space Launch System and *Orion* spacecraft will carry astronauts out of low Earth orbit, around the Moon, to the asteroids, and Mars.

15. Are there Earth jobs similar to those astronauts take on in space?

Astronauts must learn a variety of skills and do all of them well to complete a successful spaceflight. Earth professions that resemble space jobs are: research scientist, laboratory technician, pilot, scuba diver, crane operator, construction worker, emergency medical technician, computer technician, photographer, handyman, public speaker, plumber, and electrician. On every mission I sharpened old skills and learned new ones. It was the variety of jobs I trained for and executed in space that made astronaut work so interesting.

CHAPTER VIII

Walking in Space

A NASA astronaut works outside the International Space Station, 240 miles above the Earth. (NASA)

1. What would happen if an astronaut became untethered from the ship and drifted away?

When NASA's astronauts go outside the International Space Station on a spacewalk they wear a rescue system called a Simplified Aid for EVA Rescue, or SAFER. This compact jetpack is attached to the bottom of the spacesuit's life support backpack, and is designed to enable a drifting astronaut to fly back to the station.

SAFER has 24 small thrusters that shoot out cold nitrogen gas in response to commands from a small joystick. If adrift, an astronaut places the joy stick on the front of his spacesuit and turns on the jetpack. SAFER's battery-powered computer will automatically stabilize the astronaut, who then fires short thruster bursts to aim for and fly back to the station. Once there, the astronaut grabs a handrail or part of the ISS exterior, hooks on a tether, and makes his way back to the airlock.

SAFER has a limited gas supply—you might get just one chance to reach that handrail—so astronauts train frequently in a virtual reality simulator to master the flying skills needed to get back to safety. Although tested in space, SAFER has never been used to rescue an astronaut in orbit.

2. What happens if there is a leak in a spacesuit?

A leak is a truly serious emergency. A drop in oxygen pressure inside the suit can cause an astronaut to lose consciousness and die. Any large drop in pressure from a leak would trigger an alarm, and the astronaut would immediately head back to the airlock and safety.

While en route to the airlock, the suit's life support system would feed in fresh oxygen to replace lost gas and maintain pressure. The backpack's emergency oxygen tanks can replace oxygen leaking from a hole the width of a pencil lead for about 30 minutes, enough time to reach the airlock, attach a station oxygen line to the spacesuit, and get help from crewmates, ending the emergency.

Engineers design spaceship exteriors to minimize sharp points, edges, or snags, to prevent puncturing the suit's pressure bladder.

Flying without a tether, NASA astronaut Mark Lee pilots the SAFER jetpack during its test flight from shuttle Discovery on STS-64. (NASA)

3. Does the view from a spacesuit match that from the windows of the International Space Station?

The visor of a spacesuit helmet curves around an astronaut's face and allows for superb views of the Earth. I felt like I was a part of the scenery instead of an observer looking out at it. However, busy spacewalkers seldom have time to fully enjoy the view. On my spacewalks, getting five minutes between jobs to take in the view was a rare and memorable opportunity.

Inside the ISS, the best views are from the cupola on the *Tranquility* module. Seven windows provide astronauts with a glorious, bay-window view of Earth. Here ISS crewmembers use a variety of 35mm digital cameras, assorted lenses, and video cameras to capture the best Earth vistas.

4. Why is a spacewalk called an EVA?

NASA needed a term to refer to the Moon walks planned for the Apollo program, when astronauts would venture outside their spacecraft to explore the lunar surface. They coined the phrase "Extravehicular Activity," meaning activity outside the vehicle.

NASA engineers love to invent descriptive technical phrases and then condense them to abbreviations or acronyms to save time in meetings and when writing, so the term was quickly shortened to EVA (pronounced ee-vee-ay).

EVA became widely used throughout NASA and then by reporters writing stories about the Space Race. "Spacewalk" has fewer syllables than EVA, but typing only three letters is quicker.

5. Why is a spacesuit needed?

Space is a vacuum, which means there are only a few air molecules whizzing around at the altitude of the International Space Station or in deep space. Without air pressure provided by a suit, your breath would explode from your lungs, and lack of oxygen would cause you to lose consciousness in just a few seconds. Death from oxygen starvation would quickly follow. Body fluids would froth and boil as dissolved gases escaped from your lungs and bloodstream. Oh, and you'd be broiled by the Sun's heat or frozen at night by the cold of empty space. On a spacewalk, a spacesuit is the difference between life and death!

6. What are the major parts of the NASA spacesuit used at the International Space Station?

NASA astronauts use the Extravehicular Mobility Unit, or EMU, as their spacesuit. The major components are: the life support backpack (called the Primary Life Support System), the hard upper torso, the upper arm assemblies, the gloves, a chest-mounted display and control module, the lower torso assembly, an in-suit drink bag, a communications carrier assembly (headphones and microphone), a helmet assembly, a liquid cooling and ventilation garment, and a maximum absorption garment resembling an adult diaper. Russian cosmonauts use a similar spacesuit—the Orlan spacesuit—to work outside the station.

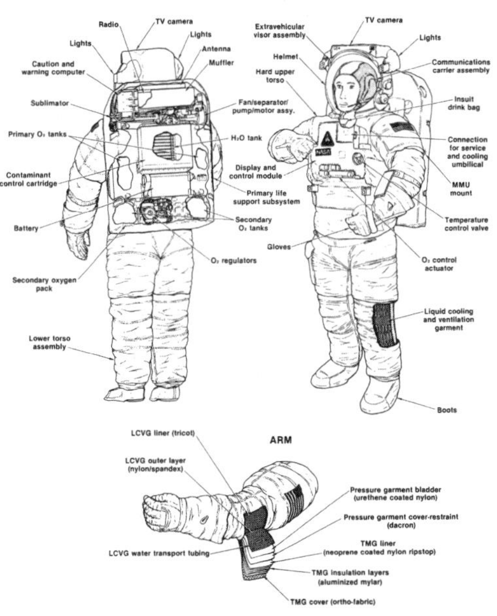

Major components of the NASA spacesuit used at the ISS. (NASA)

7. How does the spacesuit protect an astronaut?

The multiple layers of a spacesuit insulate astronauts from the harsh space environment. Next to the skin, an astronaut wears a diaper, long-john underwear, a pair of cotton socks, and thin silk gloves. Worn over the underwear, a liquid cooling and ventilation garment (LCVG) circulates water through tubes to cool the astronaut.

The arms and legs of the suit have 14 layers. The LCVG makes up the first three layers. Then there is a layer of urethane-coated nylon oxford fabric that forms the pressure garment bladder, which keeps the oxygen from leaking out and prevents lethal exposure to the vacuum of space. Next outward is a Dacron pressure garment cover which prevents the pressure bladder from ballooning outward. A neoprene-coated layer of ripstop nylon protects the vital pressure bladder from chafing, which could result in a tear. Next, five-to-seven layers of aluminized Mylar (foil-coated plastic) keep the Sun's heat out in the daytime and keep body warmth in during the frigid orbital nights.

The outermost layer of the suit is a tough Thermal Micrometeoroid Garment (TMG), made of Ortho/KEVLAR® reinforced with GORE-TEX®. It resists fire, tears, and punctures; and it protects astronauts from small pieces of space debris or meteoroids. Finally, a thin layer of gold applied to the helmet visor reflects 60 percent of the Sun's intense light and 99 percent of its heat back into space. Battery-powered heaters warm the thin silicone finger pads of the gloves, protecting the fingertips from freezing-cold metal tools and spacecraft structures.

8. What supplies and systems does the life support backpack contain?

The Primary Life Support System or PLSS (pronounced "pliss") can keep an astronaut alive for roughly 10 hours. The PLSS has a main, high-pressure oxygen tank and an emergency oxygen supply.

To cool the astronaut, pressurized tanks push water into a porous metal block where it freezes into a bed of ice when exposed to the vacuum outside. A separate water supply is pumped through tubes in this ice bed and circulates through thin tubes lining the astronaut's liquid cooling and ventilation garment.

A metal oxide cartridge in the PLSS chemically absorbs the astronaut's exhaled carbon dioxide. A rechargeable battery runs the water pump, air circulation fan, radio, and a caution and warning system that detects suit malfunctions. Separate battery packs run the glove heaters and video cameras. The oxygen and water supplies and the battery can be replenished by plugging into the station's resources in the airlock. The backpack also supports the SAFER self-rescue jetpack in case an astronaut drifts away from the station.

9. Why do astronauts have to do spacewalks?

Today astronauts perform spacewalks to maintain and repair the space station, and they can make emergency repairs on a spacecraft.

Astronauts performed more than 160 spacewalks to construct the ISS. Teams of shuttle astronauts also did EVAs, or spacewalks, to fix and upgrade the *Hubble Space Telescope*. Once, three shuttle astronauts suited up to capture a crippled satellite by hand.

Moonwalks were how the Apollo astronauts, trained as field geologists, explored the surface of another world. Apollo astronauts completed 14 moonwalks during the six lunar landing missions. Future astronauts will again do EVAs to help explore the Moon, nearby asteroids, and ultimately Mars.

10. How did you move around outside?

Spacewalking is like a ballet, performed on your fingertips. The free-fall conditions in orbit, where objects behave as if they are weightless, enable you to move your bulky, clumsy suit with great efficiency—if you do it right.

I oriented my body to keep my spacesuit and tools away from the spacecraft to avoid getting snagged on handrails and hardware. I used my hands to slowly and deliberately walk from handrail to handrail. Spacewalkers use a very light grip to conserve the strength in hands, arms, and shoulders, as those muscles are used continually for 6 or 7 hours of work outside.

Every motion in free fall, as Isaac Newton's third law says, will create an equal motion in the opposite direction. Turning a wrench without anchoring your suit will twist your body in the opposite direction. Pushing a button with your finger will push you away with equal force. Moving slowly and deliberately, using tethers and handrails, is the key to keeping your spacesuit under control.

I wore a suit that had 390 pounds (177 kilograms) of mass, and Newton's first law says that once you get that mass moving, it wants to keep right on going. So I had to be careful not to move too quickly or I'd build up too much momentum and work too hard trying to stop again. With a little practice, I was able to control my suit's motion and not waste muscle energy while getting the work done outside.

By far the easiest and most enjoyable way to move around outside the ISS was to hitch a ride on the robot arm. Stepping into a foot restraint, I could relax and enjoy a sublime view as the arm operator moved me to my next job site.

Spacewalker Steve Robinson rides the station robot arm at the ISS. (NASA)

11. Why don't astronauts fall off the International Space Station while on a spacewalk?

Like the station, spacewalking astronauts are in free fall around the Earth. Both station and astronaut are falling together at the same speed under the sole influence of Earth's gravity.

Even though an astronaut and the ISS are traveling together at a speed of five miles per second, there are no other large forces present that would cause the astronaut to separate from the station. A spacewalker is essentially falling with his home in space, and even if he or she lets go, they will keep falling right alongside each other.

I felt no sensation of falling while working outside the ISS; it felt like I was crawling around on the surface of a very solid, stable ship cruising around Earth. Though I could see the Earth moving along beneath me, I felt no sensation of speed. In a vacuum, there is no sound of wind rushing by, and no turbulence caused by the station's passage through space. My brain knew I was moving at 17,500 miles per hour (28,350 kilometers per hour), but eyes and ears told me I was gliding silently along with the station held in my fingertips.

12. What kind of tools do astronauts carry on a spacewalk?

Astronauts carry specially designed versions of familiar tools to complete repair or construction tasks outside the International Space Station. The tool handles are larger than those on Earth, so they can be grasped easily by gloved astronaut hands. Power tools are designed with large triggers and easy-to-read displays. To prevent tools from drifting away, every tool has a tether loop or eyelet built in so it can be tethered to a spacesuit.

My equipment list included waist tethers; a mini-workstation, or tool rack, clipped to my suit's chest; and a body restraint tether, like a flexible third arm ending in a gripping clamp, fastened to my left hip. I used this snake-like arm to anchor my suit to a handrail, or carry bulky equipment out to a work site.

Other tools carried on my suit included: retractable tethers and wrist tethers (to corral tools and station equipment); copper wire ties (oversize twist ties used to hold wiring in place); a Nikon 35mm camera; a battery-powered pistol-grip tool, or PGT, for driving or loosening bolts; a caddy for carrying PGT socket heads; a trash bag; and a long fabric strap with about half a dozen tether hooks attached for carrying construction equipment—we called this a fish stringer. I looked like a traveling hardware store out there!

13. Can astronauts spacewalk at night?

Roughly half of every orbit is in Earth's shadow, so spacewalkers must be able to work in darkness. The International Space Station has no external floodlights. Instead, astronauts use a pair of swiveling halogen work lamps mounted on either side of their helmet. A pushbutton on top of each lamp turns it on and off and selects either a tightly focused beam or a wider floodlight. The helmet light batteries are removed and recharged after each spacewalk.

Future moonwalkers working during the lunar night will often be able to work in Earthshine, sunlight reflected from the home planet onto the Moon's surface. A full Earth seen from the Moon is ninety times brighter than a full Moon seen from Earth. Astronauts exploring asteroids will need capable helmet lights because night on a slowly rotating near-Earth asteroid may put a spacewalker in inky darkness for hours on end.

14. Is it fun working outside on a spacewalk?

After nearly 11 years of spacewalk training, including more than 300 hours spent underwater learning how to move and manage my spacesuit and tools, I was eager to experience the real thing.

My first spacewalk was a strange and delightful physical experience. I was piloting my own little spacecraft—my spacesuit—around the outside of the station! It was a relief to learn that most tasks were easier to perform in orbit than during underwater training. A job that had me standing painfully on my head for 15 minutes during my underwater training required just five minutes of easy maneuvering while floating in free fall.

The view through my helmet visor was captivating—the starkly lit spacecraft, golden solar arrays, dazzling white spacesuits, and the always-stupendous view of Earth.

15. Is it hot or cold inside the spacesuit?

It's surprisingly comfortable inside a spacesuit. The temperature is controlled by a capable cooling system built into the suit's backpack. Without it, the body heat generated by working in the well-insulated suit would quickly overwhelm a spacewalker.

To keep an astronaut cool, the backpack circulates chilled water through the liquid cooling and ventilation garment that crewmembers wear. Astronauts can control the water temperature by turning a knob on the suit's chest-mounted control unit.

Sunlight heats the outside of the suit to 250° F (121° C). I remember exiting the airlock from *Atlantis* and feeling the Sun's heat immediately warm my arms and legs. At night, though, in the chill of space, heat slowly leaks out of the suit, particularly through the boot soles. Wearing thick wool socks and temporarily shutting off the suit's cooling system kept my feet comfortably warm.

16. Can astronauts put on and take off their spacesuits unaided?

The current suit on the International Space Station is designed to be put on and taken off singlehandedly. But it takes a lot less time if you have a friend helping you. With assistance, the suit can be put on—start to finish—in less than 30 minutes. It takes longer than that to check and double-check every connection and seal—their correct functioning is a matter of life and death.

Future suit concepts make it easier to get into and out of the suit by using a rear entry door under a hinged backpack, improving on a similar feature in the Russian Orlan spacesuit.

17. Do spacesuits fit all sizes of astronauts?

The NASA spacesuits on the International Space Station are designed to fit most astronauts. The major pieces can be mixed and matched to fit a spectrum of astronaut sizes. There are two sizes of the hard upper torso, and the legs and arms can be shortened or lengthened by inserting or removing sizing rings. Foam inserts enable smaller feet to fit snugly into boots. Glove fit is critical, so finger length and palm width can be adjusted to fit a range of hands.

NASA's shortest astronauts have trouble getting their shoulders to line up properly with the arm bearings in the upper torso. The tallest astronauts find their shoulders rub painfully against the top of the upper torso. Spacesuit designers at NASA are developing new suits that adjust to fit a wider range of astronauts while keeping the number of differently sized suit parts to a minimum.

18. How do astronauts eat and drink inside a spacesuit?

Spacewalkers sip from a 2-liter water bag that is attached by Velcro to the inside of the front chest area of the spacesuit. A drinking tube with a valve comes into the helmet just above the neck ring, close to the astronaut's mouth. Biting the valve allows the astronaut to suck water from the tube.

Early space shuttle spacewalkers snacked on food sticks—thin, chewy fruit bars wrapped in edible rice paper—mounted on the neck ring of the suit. An astronaut could duck his head down to take a bite and, at the same time, slide the fruit bar up for the next nibble. The fruit bar tasted OK, but it left a sticky residue on the face, so it just wasn't worth the trouble. Astronauts have since decided it's best to have a big breakfast before the spacewalk and then enjoy a satisfying supper afterward.

19. Can you scratch your nose inside a spacesuit?

The spacesuit fits very snugly, so it's impossible to pull your hand out of the glove and arm of the suit and reach up to scratch your nose inside the sealed helmet. I did stick a small, one-inch foam cube down on the helmet neck ring to the left of my chin. I pinched my nostrils against this stiff foam to clear my sinuses when the pressure changed inside the suit, but it was also handy for scratching my nose and chin.

Unfortunately it was too far down in the helmet to rub a brow, or help absorb tears from a watery eye.

20. How do astronauts go to the bathroom when wearing a spacesuit?

On a spacewalk, time is so precious that astronauts can't afford to spend several hours coming back inside to undress, just to use the bathroom. So spacewalkers wear adult diapers under their water-cooled, long-john underwear. In free fall conditions the diapers are essential to absorb liquid urine that would otherwise migrate everywhere and contaminate the suit. Remember, everything floats in free fall, so without a diaper things could get messy! I didn't think twice about wearing one. After the spacewalk, I would seal the diaper in a Ziploc bag and toss it in the trash. I then wiped down the interior of the suit to eliminate bacteria. It's hard to imagine shipping diapers to the Red Planet for use on Mars walks, but a better solution has yet to be found.

21. What kind of air is in the spacesuit?

In contrast to the station's atmosphere of 80 percent nitrogen and 20 percent oxygen, we use pure oxygen inside a spacesuit. You don't need nitrogen to stay alive, and breathing oxygen at a pressure of about 4.3 pounds per square inch (psi) makes your red blood cells just as happy as if they were at sea level on Earth. So breathing pure oxygen not only keeps you alive, it also enables us to lower the pressure inside the suit, making it more flexible and less like a stiff, inflated balloon.

Before lowering the pressure inside the suit to 4.3 psi, astronauts must breathe pure oxygen in the suit for at least 90 minutes prior to going outside. This "pre-breathe" removes dissolved nitrogen gas from the blood, which greatly reduces the possibility of painful decompression sickness caused by nitrogen bubbles forming in the blood as suit pressure drops.

After 8 to 10 hours breathing pure oxygen during their spacewalk, astronauts come back inside the ISS, remove their helmets, and begin breathing nitrogen and oxygen again until their next EVA.

22. Was it quiet or noisy inside the spacesuit?

The only noise generated by my spacesuit came from a small, high-speed fan in the backpack that blew fresh oxygen into the back of my helmet. Just before a spacewalk, as the oxygen pressure decreased inside the suit, the fan noise faded away. With fewer oxygen molecules inside the helmet to carry the sound, the fan noise dropped to a whisper. As the pressure dropped further, the pitch of my voice changed to a lower, "fuzzier" note. No sound carried through the vacuum outside, but if I banged a tool on the suit's metal fittings, I could hear that "thunk" clearly inside. On my spacewalks, there were long periods during which the only sound I heard was that of my own breathing. Only the radioed voices of my crewmates through my headphones broke the silence.

23. What does wearing the suit feel like?

The closest thing to wearing a spacesuit that I've experienced is scuba diving in a wet suit. The spacesuit is far more comfortable to wear, however. My head was free to move, my vision was unrestricted, and I didn't have to breathe through a mouthpiece.

After spending hundreds of hours in a suit during spacewalk training, it felt very much like a second skin. I was comfortable enough while working outside that I barely noticed I was wearing it. Instead, I focused completely on my hands, tools, and the job I had to do.

Occasionally, I got a finger pinched in a glove or chafed shoulders from rubbing against the suit's interior. Making sure the gloves fit well and putting moleskin on pressure points took care of most rough spots.

24. What keeps astronauts and tools from floating away from the International Space Station?

Tethers are hard to see in television shots of spacewalking astronauts, but they are what keep astronauts and all their equipment attached to the spacecraft or spacesuit, preventing them from drifting away.

Astronauts tether themselves to the ISS with a retractable, 50-foot stainless steel cable. Its reel is attached at the suit hip and the other end

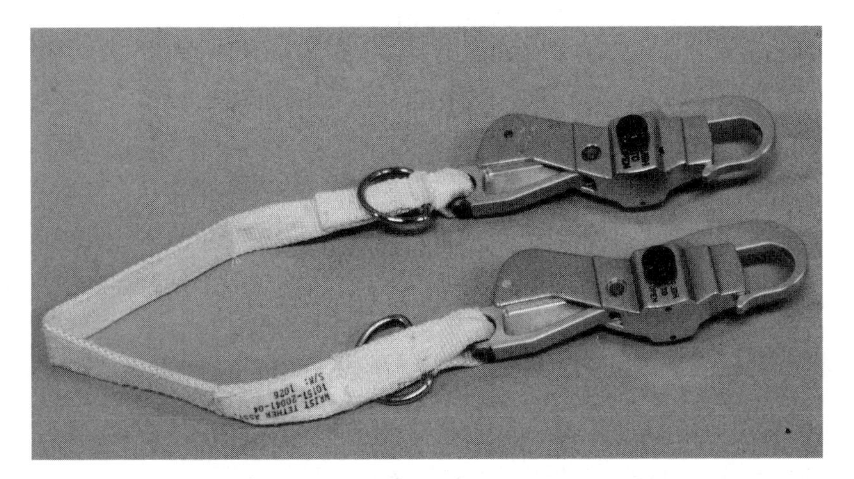

A Kevlar EVA wrist tether, used to anchor tools to a spacesuit or the ISS. (NASA)

is clipped to a handrail on the station. They also use 3-foot (1-meter) Kevlar® tether straps to anchor themselves to a station worksite.

Tools are locked firmly to a tool rack worn over the chest of the spacesuit. Additional tools are carried on short, high-strength Kevlar® straps, or are hooked to retractable equipment tethers.

25. Do astronauts lose things outside?

The most important rule to remember to prevent losing equipment outside is: always tether! Never move any tool or equipment unless you tether to it first. Don't remove the tether until the tool is safely locked down or put away.

Still, mechanical hooks on our tethers do fail, and astronauts sometimes forget to tether, or they bump into equipment and cause locks or tethers to open. On the first U.S. spacewalk in 1965, a loose outer glove drifted away from the open hatch of *Gemini IV*. Astronauts at the station have lost a wire spool, a tool bag, a foot platform, and assorted other tools.

During my spacewalks with partner Bob Curbeam, we never lost an item. But once, a caddy holding some of our socket heads did start to drift away when its tether hook stuck open slightly. Bob spotted it floating lazily about ten feet away and managed to recapture it. A good astronaut habit is to tug on every tether to make sure the tool is firmly attached and the hook closed.

26. What happens if a tool or other equipment is lost on a spacewalk?

We have spares aboard the station for some critical EVA tools in case we lose them. Eventually a drifting tool will be pulled to a lower orbit by atmospheric drag and burn up in the atmosphere. If an essential item goes missing, engineers will send a replacement in the next cargo shipment.

27. How much does a spacesuit at the International Space Station weigh?

NASA's ISS suits weigh in at just over 390 pounds (177 kilograms)—empty. With an astronaut in it, we're talking about 550 pounds (250 kilograms). Work outside also requires adding up to 40 pounds (18 kilograms) of tools and equipment on hip brackets or a chest rack. Now, in orbit, it's possible to lift that bulky suit with a finger, but its mass, or inertia, makes it hard to get moving and to stop. Think of a spacewalk as being inside a refrigerator, sticking your arms out its sides, and then maneuvering that mass around the outside of the space station.

28. How do astronauts train for the work they do on a spacewalk?

The best way we've found to reproduce the spacewalking environment is to train underwater in a neutral buoyancy tank. NASA astronauts train in the 40-foot-deep, 6-million-gallon pool of the Neutral Buoyancy Lab (NBL) in Houston, Texas.

Divers add or remove weights from an astronaut's suit to keep it from sinking to the bottom or floating to the surface. Astronauts are suspended underwater and float there just like they would in space. In this simulated weightless environment, astronauts learn how to handle their tools, practice attaching and detaching tethers, and how to move the suit efficiently without wasting muscle power.

Working on submerged, life-sized mockups of the station and other spacecraft, astronauts also practice repairs and maintenance work they might be assigned.

In advance of an orbital spacewalk, Mission Control sends up video lessons to the station so astronauts can review upcoming EVA procedures before their next trip outside.

Astronauts Stan Love and Stephen Bowen practice asteroid exploration techniques in the Neutral Buoyancy Lab at Johnson Space Center. (NASA)

29. Is spacewalking more dangerous than other mission tasks?

Just living on the International Space Station exposes one to risk from solar and cosmic radiation and from collisions with micrometeoroid or orbital debris which can puncture a module and cause a sudden loss of cabin pressure. But on an EVA, a spacewalker deliberately leaves the protective structure of the spacecraft and relies solely on the thin fabric and the small, pressurized volume of the spacesuit. Fabric is not as tough as metal, and a puncture will empty your spacesuit more rapidly than a much bigger station module. A spacesuit also provides less radiation and thermal protection than a spacecraft, although exposure times outside are short.

Another increased risk is that spacewalkers are farther from a safe haven. In an emergency, it may take astronauts tens of minutes to return to the airlock and get help from their crewmates.

30. Is there a danger of meteoroid strikes while outside on a spacewalk?

Natural debris from asteroids and comets and bits of man-made space junk are constantly speeding through space, and could harm a spacewalking astronaut. Some protection comes from working next to the large modules of the International Space Station, which help shield astronauts from debris strikes. The outer layer of the spacesuit, woven from Kevlar®, also protects from penetrations by debris particles.

So far, neither micrometeoroid nor man-made debris particles have penetrated a spacesuit or its life support system. However, tiny, sharp-edged craters caused by impacts on ISS handrails have sliced into the silicone rubber coating on astronaut gloves, creating the risk of a pressure leak. Those gloves had to be returned to Earth for repair.

Micrometeoroid or space debris impact damage to one of the solar panels at the ISS. (NASA)

31. What failures have astronauts experienced in the spacesuit?

In July 2013, European Space Agency astronaut Luca Parmitano was wearing a U.S. spacesuit on a spacewalk to repair the International Space Station when water began leaking into the duct feeding oxygen into his helmet, near the base of his neck. Water droplets coalesced into a thin film of water on the back of his communications cap and began to spread and creep toward his face. He reported the wet feeling to Mission Control, but everyone—including Luca—assumed it was sweat or water leaking from the drink bag near his chin.

Finally, when the spreading blob of water began to creep into his eyes and nostrils, the flight director halted the spacewalk and directed Luca and his partner back to the airlock. Although Luca shook his head repeatedly, he couldn't fling off the water; it blinded him and filled his ears. Luca was in danger of choking, and perhaps drowning, inside his helmet.

His EVA partner followed him inside the airlock, closed the outer hatch, and the ISS crew inside re-pressurized the chamber. They quickly removed Luca's helmet and toweled the water from his face, ending the emergency. Engineers diagnosed the problem as a clogged backpack water separator, caused by mineral impurities in the station's water supply. They later sent up spare parts to repair the suit.

32. What work did you do on your spacewalks?

My spacewalking partner, Bob Curbeam, and I performed three spacewalks to help build the ISS. We prepared a docking port to host the new *Destiny* laboratory; connected electrical power lines, data cables, and the station's cooling lines to the lab; and equipped the lab exterior for future expansion of the station. We unveiled the lab's new Earth observation window, and then connected the station's second docking port to the front of the lab. We installed a refrigerator-sized spare radio transmitter on the station's truss section, and finally tested ways to move an injured or unconscious crewmember back inside the airlock.

The author working outside the International Space Station on space shuttle mission STS-98. (NASA)

33. How long were you outside the space shuttle and the International Space Station?

During three spacewalks, I spent 19 hours, 49 minutes working outside the ISS with Bob. On our longest EVA we were outside for 7 hours, 34 minutes.

34. Why is spacewalking a challenge?

The many protective layers of a spacewalk suit make the arms and legs bulky and difficult to move. Filling the suit with oxygen, even at low pressure, adds to this stiffness. The thick, pressurized gloves require a near-constant squeezing of the fingers to grip handrails and tools. The steady demand on upper body, shoulder, and forearm muscles tires them and tends to overheat the astronaut. The exacting nature of the work you do on a spacewalk, where mistakes can have deadly consequences, also calls for extraordinary concentration. All these factors combine to make a spacewalk one of the most arduous, yet wondrous and rewarding astronaut jobs.

Astronaut Bob Curbeam takes in the view outside the station's Destiny lab. (NASA)

35. How did astronauts move around on the Moon?

The Apollo astronauts reported that the easiest way to get around in the Moon's weak gravity (one-sixth that of Earth) was by using a two-footed bounce or skip, much like a kangaroo hop.

Moonwalkers could go faster by loping along, pushing off from one foot to the other. But bounding along too fast in the weak lunar gravity made it hard to stop, resulting in a few falls that showered dust over their suits and equipment.

36. What new spacesuits is NASA designing for future exploration?

NASA has begun developing new spacesuits for deep-space exploration in the decades ahead. Astronauts on the new *Orion* spacecraft will at first use the shuttle's Modified Advanced Crew Escape Suit (MACES), for launch and reentry. The MACES can also be used for spacewalks to make emergency repairs, but it's not designed for extended work outside.

A Modified Advanced Crew Escape Suit with an advanced life support backpack and better gloves is being studied by NASA for a 2020s

Apollo 17 astronaut Jack Schmitt lopes across the Moon's surface. (NASA)

visit to a boulder captured from an asteroid. The agency is also developing advanced exploration suits for eventual expeditions to the Moon or asteroids.

These new spacesuit prototypes are intended for work in a vacuum and low gravity (left), and on the surfaces of the Moon or Mars (right). (NASA)

37. What changes will spacesuits need for use on asteroids or other planets?

These new suits will have smaller, more efficient life support systems, powered by lighter, longer-lasting batteries. Storing oxygen in liquid form instead of as a high-pressure gas will also increase the amount of time an astronaut can spend outside on an asteroid or planet.

The suit joints will have to be more flexible so astronauts can walk, bend, and kneel on planetary surfaces. More flexible gloves will increase dexterity and reduce hand and arm fatigue. Advanced computers built into the suits will enable checklists, images, and sensor readings to be projected on the faceplate inside the astronaut's helmet. These improvements will be tested on the ISS and on asteroids and the Moon, and will lead to a spacesuit worn by astronauts on the first expeditions to Mars.

CHAPTER IX

Risks of Space

Shuttle Challenger lifts off on its final mission, STS-51L, on January 28, 1986. Ship and crew were lost less than two minutes later. (NASA)

1. What are the biggest risks faced by astronauts in space?

Launch failures are a serious risk. Developing a reliable rocket booster that can overcome Earth's gravity and make it safely into space is a daunting engineering challenge. To enable a crew to get away from a failing rocket and land safely, a launch escape system is essential.

Once in space, astronauts face other risks. Space debris and micrometeoroids can puncture the spacecraft and allow the air to escape. A fire in the confined spacecraft cabin can consume all the oxygen or expose the crew to toxic smoke. Toxic gases from cooling systems can leak into the cabin or contaminate a spacesuit. A failed docking maneuver can cause a collision between spacecraft and rupture the cabin. On long expeditions into deep space, radiation and the harmful effects of extended life in free fall can damage an astronaut's health.

Lastly, a heat shield failure while coming back through the atmosphere at speeds of 17,500 miles per hour (28,350 kilometers per hour) or more could incinerate the spacecraft. That heat shield must work every time it's used.

2. What is space junk, and is it a hazard?

Low Earth orbit is strewn with everything from spent rocket boosters to dead satellites to small pieces of shattered spacecraft, all circling the globe at collision speeds of tens of thousands of miles per hour.

The International Space Station is routinely hit by small pieces of debris, pitting its windows and damaging solar panels and handrails used by spacewalkers. Larger strikes could puncture the hull or damage critical systems. The good news is that no major damage to the ISS has occurred so far, and there's no chance of a massive space debris storm like the ones you see in the movies. However, strikes from larger pieces of debris are still a serious risk.

The Air Force's Space Surveillance Network uses radar to track debris fragments larger than a baseball. If the network shows debris getting within about 2 kilometers (a little over a mile) of the ISS, Mission Control can maneuver the station slightly to avoid a collision. If a warning comes too late for a rocket maneuver, the astronauts may be asked to retreat to the *Soyuz* or U.S. "lifeboat" capsules until the danger has passed.

Space junk will be a hazard for decades. Space-faring nations must ensure their satellites and rocket booster stages don't contaminate near-Earth space. But until nations stop generating debris and the threat recedes, new spacecraft must be equipped with effective debris protection.

3. Are micrometeoroids a serious hazard?

Impacts from micrometeoroids, which are tiny pieces of asteroids and comets, pose a danger to spacecraft flying above the Earth's atmosphere. A high-speed impact could puncture the ship like a bullet, letting cabin air escape or disabling critical systems like engines or life support. The risk is real, but space is a big and mostly empty place, and the chance of a ship or station being struck is relatively small.

To prevent damage to the International Space Station, which must survive in orbit for about 25 years, its habitable modules are armored with debris shields. These are thin metal and fabric layers that can break up and slow down pieces before they penetrate the hull. The ISS debris shields can probably stop fragments up to the size of a pea.

4. How would astronauts fix the damage from a space debris strike?

If a piece of space debris collides with the International Space Station and punctures a living module or laboratory, the air inside would begin leaking into the vacuum of space. The crew would work quickly to try to pinpoint the leak and isolate the damaged module. If the hole is visible and not hidden behind a system or experiment rack, the crew is trained to patch it before pressure levels drop dangerously.

If they can't find the hole, the astronauts will evacuate the damaged module and close hatches to stop air from being lost from the rest of the station. With the damage now confined, the crew might do a spacewalk to patch the hole from outside, then reopen and repair the module. If the station is too severely damaged, the crew can escape to Earth using their docked lifeboat spacecraft.

5. How do astronauts know if there is a fire or other serious problem aboard the International Space Station?

Computer systems aboard the ISS monitor life support functions and spacecraft performance, as do flight controllers in Mission Control. In the case of a serious emergency, such as a fire, the ISS caution and warning system sounds a loud, piercing alarm and a warning light appears on each module's caution and warning panel.

Class 1 alarms—fire, rapid depressurization, and toxic atmosphere—are the most serious emergencies. These are emergencies that could prove life-threatening without *immediate action* by the crew and Mission Control. Less serious alarms are called Warnings, Cautions, and Advisories. They still require a quick response, but trigger different tones and warning messages on the crew's laptop computers.

Each month, ISS crews hold emergency drills to rehearse the critical steps to be taken by each crewmember should a serious problem occur.

Astronauts Terry Virts, left, and Samantha Cristoforetti practice emergency procedures aboard the International Space Station in the Zarya module. (NASA)

6. How does fire behave in a spacecraft?

Fire in a spacecraft behaves differently than on Earth. In free fall, the heated air from a flame does not rise. Instead, a layer of burned hot air—now low in oxygen—remains right around the flame. Without a fresh supply of oxygen, the fire is slow to spread. The flame resembles a glowing ball, burning at lower temperatures than flames on Earth. NASA is investigating how fires burn in space using the combustion research facility in the station's *Destiny* lab.

If a fire occurred on the ISS, the crew would first put on breathing masks and grab flashlights and the emergency checklist. They would then go to the nearest laptop control computers to try to locate the fire and remove electrical power that might be its source. Other crewmembers would try to put out the fire using the extinguisher in the affected module. If the fire is uncontrollable, the crew would leave the module, seal its hatches, and shut down air circulation fans to prevent heat and toxic smoke from spreading to the rest of the station.

The firefighting steps on the space station were developed based on lessons learned from a dangerous 1997 fire aboard the *Mir* space station,

A candle flame burning on Earth (left) and in free fall conditions (right). Flames in free fall take on a spherical shape because air heated by the flame doesn't rise. (NASA)

when a burning oxygen generator spewed molten metal and toxic smoke throughout the spacecraft.

7. Is radiation a big problem for astronauts?

Astronauts on the International Space Station are protected from most solar and cosmic ray radiation by the Earth's geomagnetic field. Further protection is provided by the station's metal hull and a layer of radiation-absorbing plastic that lines the crew's sleeping quarters.

However, in deep space the spacecraft hull alone won't provide enough protection from solar storms or cosmic rays. Unprotected astronauts may absorb enough radiation to increase their cancer risk significantly, or even suffer radiation sickness during a fierce solar storm. Such a storm might force the crew to retreat for several days into a small, water-lined storm shelter until radiation levels subside.

Dealing with cosmic rays will require more massive shielding (liquid hydrogen or water is most effective), or nuclear-powered rockets to get space explorers to their destinations faster to reduce the duration of radiation exposure. On the Moon or Mars, explorers will have to live underground to avoid hazardous radiation. Good locations for a base may be in lava tubes, or in craters, shielded by dirt bulldozed over the living modules.

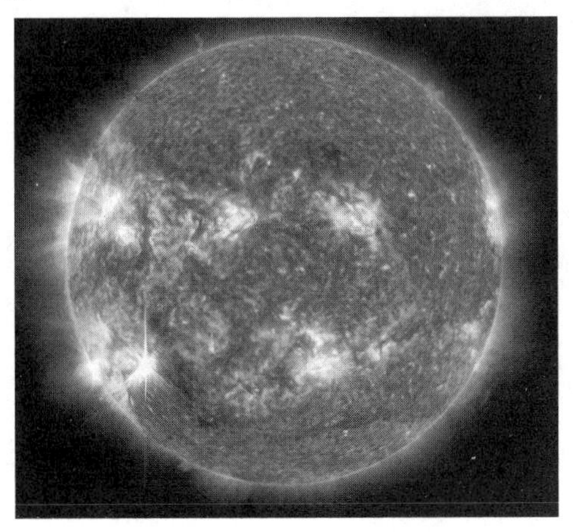

Solar flares generate streams of charged protons and electrons, harmful to astronauts in deep space. (NASA)

8. Will life in free fall prove to be a health hazard?

Aerobic exercise on a treadmill, cycle ergometer, or rowing machine largely protects muscles, the heart, and lungs from the "laziness" effects of living in free fall. Astronauts have found that the ISS Advanced Resistive Exercise Device, a strength-training machine, seems to be effective at reducing the slow bone calcium loss experienced during free fall.

Prolonged stays in free fall can cause changes in an astronaut's vision, related to increased pressure inside the brain. This is apparently due to the shift of body fluids "floating" from the lower body up into the chest and head. Another worry is that living in free fall weakens the immune system, reducing an astronaut's ability to fight off germs and infections.

If the negative effects of free fall can't be overcome, engineers can design a spacecraft that rotates the living spaces like a centrifuge to provide the crew with artificial gravity. But such a system increases the complexity, weight, and cost of the spacecraft, so space program managers are eager to avoid that option.

The Earth-orbiting, rotating space station providing artificial gravity in the movie 2001: A Space Odyssey. (Image by Nick Stevens, International Association of Astronomical Artists)

9. How would astronauts respond to a medical emergency in space?

To help a sick astronaut, the International Space Station has a well-equipped medical locker with trauma gear and medicines to treat almost every malady. A flight surgeon is also as close as the radio, ready to provide advice.

All ISS astronauts get medical training to equip them with the skills needed to handle medical emergencies, rather like an emergency medical technician (EMT) on Earth. The good news is that cold and flu germs have a hard time making it up to orbit to cause illness. But in a life-threatening medical emergency, such as a heart attack or appendicitis, the crew would stabilize the sick astronaut and return him quickly to expert care on Earth via the *Soyuz* or space taxi lifeboats.

On deep-space or Mars expeditions, medical experts on Earth can lend long-distance advice, but due to their distance from home, astronauts will have to treat all serious illnesses and injuries on their own. Such deep-space crews will almost certainly include a physician.

10. Can we rescue astronauts in space?

The International Space Station has a built-in rescue system—two Russian *Soyuz* space lifeboats that are always docked at the station. If a serious emergency forced the crew to abandon their outpost, they would scramble into their *Soyuz* lifeboats, undock, and return to Earth and safety. Future commercial transport ships will also be used as lifeboats during their stay at the station.

If a transport spacecraft became stranded in low Earth orbit, the space-faring nations would likely team up to launch a rescue craft, like another *Soyuz* or a U.S. space taxi.

In deep space, though, both distance and the lack of capable rescue craft will make a quick return to Earth quite difficult. The best strategy is to build reliable spacecraft systems with backups, so the crew can reach a safe haven on the Moon or Mars to await rescue.

11. What fatal space accidents have occurred?

1967: Three NASA astronauts—Virgil "Gus" Grissom, Edward White, and Roger Chaffee—lost their lives in an *Apollo* spacecraft fire on the launch pad.

1967: Russian cosmonaut Vladimir Komarov perished when his *Soyuz* craft's parachute failed during reentry.

1971: Cosmonauts Georgi Dobrovolski, Vladislav Volkov, and Viktor Patsayev perished when a faulty valve depressurized their *Soyuz 11* spacecraft.

1986: The crew of shuttle *Challenger*—astronauts Francis "Dick" Scobee, Michael Smith, Judith Resnik, Ellison Onizuka, Gregory Jarvis, Ronald McNair, and Christa McAuliffe—died when a booster failure destroyed their orbiter during launch.

2003: The shuttle *Columbia* crew—astronauts Rick Husband, William McCool, Kalpana Chawla, Laurel Clark, David Brown, Michael Anderson, and Ilan Ramon—perished when a damaged heat shield caused their orbiter to break up during reentry.

12. Why was *Challenger* lost?

On the morning of January 28, 1986, NASA made the critical mistake of launching the shuttle after a night of below-freezing temperatures. The cold caused a seal in the right solid rocket booster to fail and let hot exhaust gas burn a hole through the side of the rocket casing.

About 72 seconds after liftoff, the escaping rocket exhaust caused the booster to break loose from and strike the external fuel tank. The external tank ruptured, tossing the orbiter sideways and exposing it to extreme aerodynamic (or air pressure) forces that broke up the spaceship. The astronauts perished when their cabin either lost pressure at breakup, or when it struck the water some four minutes after launch. Killed were astronauts Gregory Jarvis, Christa McAuliffe, Ronald McNair, Ellison Onizuka, Judith Resnik, Francis "Dick" Scobee, and Michael Smith.

After the accident, NASA improved the booster design and changed its engineering, communication, and management practices to improve safety.

The Kennedy Space Center's Visitor Complex houses a memorial to the two shuttle crews who gave their lives while exploring space, honoring the astronauts and displaying pieces of both Challenger and Columbia. (NASA)

13. What caused the loss of the shuttle *Columbia*?

Seventeen years after *Challenger* was lost, *Columbia* launched on a science mission (STS-107). About 82 seconds into the flight on Feb. 1, 2003, a briefcase-sized chunk of insulating foam tore off the nose of the external fuel tank and struck *Columbia's* left wing.

NASA learned of the foam impact shortly after *Columbia* reached orbit, but decided that any damage the wing sustained would not endanger the astronauts. In fact, the impact had cracked a heat shield panel on the front edge of *Columbia's* left wing.

Sixteen days later, as *Columbia* returned to Earth, the extremely hot gases that formed during reentry penetrated the wing, weakening it internally. Traveling at 12,000 miles per hour (19,440 kilometers per hour), 38 miles (62 km) above Texas, the wing failed and the orbiter tumbled out of control. The rapid breakup of the orbiter followed, causing a loss of cabin pressure and destruction of the crew cabin. Killed were astronauts Michael Anderson, David Brown, Kalpana Chawla, Laurel Clark, Rick Husband, William McCool, and Ilan Ramon.

Following the accident, NASA improved the performance of the foam insulation on the external fuel tank, started in-orbit inspections of the orbiter heat shield surfaces, developed heat shield repair tools, and created shuttle rescue plans. The agency once again revamped its management and decision-making procedures to improve flight safety.

The STS-107 Columbia crew, shown in orbit during their 2003 mission. Top row (left to right): Brown, McCool, and Anderson. Bottom row: Chawla, Husband, Clark, and Ramon. (NASA)

14. Why didn't their spacesuits save the *Columbia* astronauts on their return to Earth?

When *Columbia's* left wing buckled and broke loose during reentry, the orbiter tumbled out of control. Within seconds the crew cabin started to come apart and rapidly lost pressure. The astronauts had their helmet visors up and most had their gloves off, so their suits could not hold pressure, and they quickly lost consciousness. But even with their visors closed and gloves on, their suits were not built to withstand an encounter with the extreme aerodynamic forces as their ship came apart at 12,000 miles per hour (19,440 kilometers per hour), roughly 200,000 feet (60,960 meters) over Texas. The rapid tumbling and sudden deceleration forces they encountered were not survivable.

15. Was the movie about the *Apollo 13* space near-disaster accurate?

The visual accuracy of that 1995 film was impressive. Some of the spacecraft interior scenes showing the astronauts floating in space were filmed aboard NASA's KC-135 "Vomit Comet" aircraft, which creates 20-second periods of free fall during flight. The historical account is also quite accurate, although the filmmakers reduced the size of the ground team from hundreds to a few dozen, so they could focus on just a few key characters.

Astronauts Fred Haise and James Lovell of the *Apollo 13* crew have said that the director did exaggerate scenes showing the crew arguing during the mission as a way to heighten the dramatic tension. The *Apollo 13* filmmakers were very successful at showing how teamwork in space and on the ground transformed a near-fatal tragedy into a triumph of human determination and survival.

16. Can you look at the Sun in space?

That's not a good idea at all. Without the Earth's atmosphere to absorb and scatter the Sun's rays, its light is far too hot and intense for your eyes' tender, light-sensitive retinas. Even from inside the cabin, looking through several panes of thick glass with a heat-reflecting coating, a few seconds of staring at the Sun can burn your retinas permanently. Sunglasses would reduce the intense glare, but they still don't provide adequate protection.

The helmet astronauts wear on a spacewalk doesn't have triple-pane glass; it's just a clear plastic bubble. During daylight, spacewalkers lower an outer, polycarbonate (plastic) visor coated with a thin layer of gold. The gold lets in about 40 percent of the visible light, but reflects nearly all the heat from the Sun, protecting the eyes. The visor also screens out harmful ultraviolet light. It's not possible to safely look at the Sun, but at least astronauts can work comfortably in the brilliant sunshine.

17. Were you afraid while you were in space?

Right before launch I felt a few butterflies in my stomach, as if I was getting ready to go on stage for a big performance—and I suppose I was. I didn't worry about the safety of the shuttle: other experts had the job of making our ship ready for a safe flight. My biggest worry was whether *I* had prepared thoroughly enough for the mission, so I wouldn't make a serious mistake and let down my crew, the Mission Control team, and NASA. Once I reached orbit, I don't recall experiencing any anxiety on any of my missions. I was too busy!

Shuttle Endeavour carries the author and the crew of mission STS-68 into orbit six weeks after a launch pad abort. (NASA)

18. What new hazards will we face on the Moon, asteroids, or Mars?

Going beyond the International Space Station and leaving Earth orbit will expose astronauts to several new hazards. After they leave the protection of Earth's geomagnetic field, their spacecraft will be exposed to much higher levels of radiation. This will come from solar storms (high-speed charged particles from the Sun) and galactic cosmic rays (very high-speed, heavy atomic nuclei), both of which can sicken a crew or increase their risk of getting cancer.

Landing on a moon or planet requires powerful rockets or heat shields and parachutes, all of which must work perfectly for astronauts to reach the surface safely. While exploring our Moon or Mars, astronauts will face dangers from sharp rocks, falls, dust storms, toxic soils, and continued radiation exposure.

Astronauts on long expeditions, such as those to Mars or an asteroid, will face another hazard—loneliness. They will be so far away that normal radio conversation will be impossible, and they will feel the psychological effects of isolation from family, friends, and Earth's familiar environment.

19. How are future ships being made safer?

The latest generation of commercial space taxis and the *Orion* spacecraft being developed for deep-space journeys all incorporate launch escape systems to push or pull the crew module clear of a failing rocket booster.

Smaller, faster computers will make it easier to control spacecraft during orbital flight, rendezvous and docking with other spacecraft, and reentry to Earth's atmosphere. Automating most routine flying tasks will reduce the chance of human-caused piloting errors. Computers can also keep watch over spacecraft systems to detect failures in their earliest stages, before they become more serious.

More reliable life support systems will provide astronauts a safer and more comfortable living environment. Shielding made from sandbags filled with lunar or asteroid dirt will help protect crews from impacts and radiation hazards. Failures will still occur, of course, and one

The launch abort system pulls an unpiloted Orion spacecraft to safety in a May 2010 test flight. (NASA)

challenge in deep space will be to keep computers and machines functioning for years on end, far away from maintenance shops on Earth.

20. Will space tourism be safe?

No machine built by humans is failure-proof. Even vehicles that launch on short, suborbital tourist flights to the edge of space undergo accelerations of 4 or 5 g's (4 or 5 times the force of Earth's gravity), experience temperatures of hundreds of degrees Fahrenheit, and reach speeds of 3,000 miles per hour (4,860 kilometers per hour) or more. In 2014, a test version of the Virgin Galactic *SpaceShipTwo* commercial space plane broke up in flight, killing one of the pilots. Testing and competition should improve the safety of these spacecraft, but passengers should be fully acquainted with the risks of spaceflight before strapping in for a ride to the edge of space.

Seeing Space

NASA astronaut Terry Virts photographs Earth from the cupola aboard the International Space Station. (NASA)

1. What does an orbital sunrise look like? How many do you see every day?

The International Space Station goes around the planet every 92 minutes or so, meaning astronauts see nearly 16 sunrises and 16 sunsets every 24 hours. Sunrises begin with the appearance of a thin, indigo line along Earth's horizon, changing to a light, robin's egg blue as the ISS heads toward dawn. A rainbow of colors spreads rapidly along the planet's edge, adding red, orange, and yellow. Then there's a brilliant burst of white light as the Sun's disk crests the atmosphere. The subtly glowing horizon explodes into white-hot sunlight in just about thirty seconds. Even digital cameras have trouble capturing the delicate colors and rapidly changing light levels of an orbital sunrise.

2. What does the night sky look like from space?

From above the atmosphere, the full glory of the night sky is visible once your eyes adapt to the dark. That's a challenge, because an orbital night is at most 45 minutes long, and your eyes can take up to 30 minutes to dark-adapt.

To get the best view, after sunset astronauts must turn down all the interior lights. As their eyes adapt, they will begin to enjoy a spectacular view of the Milky Way and thousands of stars surrounding the black void of Earth's night hemisphere. So many stars are visible that astronauts have trouble identifying the familiar constellations, as their bright stars are swamped by the light from thousands of others. Looking down toward the darkened Earth, they'll catch a glimpse of an occasional meteor burning up in the atmosphere below.

3. Why don't the stars twinkle when seen from space?

Turbulence in the Earth's always-moving atmosphere causes the air to diffract, or bend, starlight in a slightly different direction every few milliseconds. To our eyes, this rapid change in position makes the star appear to be twinkling. We also call this scintillation.

A view of the night sky with our Milky Way galaxy, seen from the International Space Station. (NASA)

Above the atmosphere, astronauts and telescopes have no air between them and the starlight streaming across space, so the stars and planets shine steadily. Getting away from the atmosphere's turbulence and absorption of light is why astronomers pushed to put great observatories like the *Hubble*, *Chandra*, *Spitzer*, and *Compton* spacecraft into space. The *James Webb Space Telescope*, which will look at infrared light (we think of it as heat from stars and galaxies), will be the next great observatory to be launched above the distorting effects of our atmosphere.

4. Over what parts of Earth does the International Space Station fly?

Because the station orbits Earth in a plane inclined to the equator by 51.6°, the planet's rotation eventually causes the ISS to pass over all ground locations between 51.6° north and south latitude. Astronauts can view land and seascapes from north of the Great Lakes to the tip of South America, and from central Siberia to south of New Zealand.

With Lakes Superior and Michigan at left, portions of all five Great Lakes are seen in this northeast-looking view from shuttle Columbia on STS-40. (NASA)

5. Can you see the Great Wall of China from space?

The Great Wall is made of natural stone and mud bricks, which are largely the same color as the ground surface around the Wall. With an average width of just 16 feet (about 5 meters), the Wall is far too narrow to detect with the naked eye. But this historic structure can be detected in images taken from orbit. Using digital cameras and telephoto lenses, astronauts can take photos of the regions crossed by the Wall. Although you won't be able to see it looking out your spacecraft window, you can certainly bring back an image that, when enlarged, will show the Wall.

This space radar image from my first shuttle mission shows a section of the Great Wall in a desert region about 440 miles (730 kilometers) from Beijing. The vertical line indicates portions of the Wall dating to the fifteenth century. (NASA)

6. What evidence of human civilization can astronauts see with the naked eye?

From the International Space Station or a spacecraft in low Earth orbit, astronauts can see many traces of our civilization. The human eye is especially good at recognizing patterns and straight lines, so in daytime it's easy to see highways, railroad lines, airport runways, jet contrails, and the wakes of ships at sea. Cities appear as gray smudges on the landscape. At night, populated areas glow like diamonds on black velvet, connected by the spidery webs of highways—an eye-popping illustration of our society's reach across the planet. For many examples, see NASA's Gateway to Astronaut Photography of Earth website: eol.jsc.nasa.gov.

7. How do cities look from orbit?

By day, the roads, masonry walls, and rooftops of cities appear gray against the natural vegetation cover on Earth. Because of the haze and pollution created by densely populated cities, they are not always clearly visible. Desert cities and those surrounded by winter snow cover are far easier to spot than those situated in agricultural or forest regions. My hometown of Baltimore, Maryland is easily found from orbit by tracing the shoreline contours of the Chesapeake Bay.

8. How would you describe Earth as viewed from space?

Looking at Earth is like a geography lesson come alive! The planet is endlessly fascinating, displaying striking colors and pastel hues in a palette that is always changing, and always beautiful. Some of the distinctive sights seen from orbit include: the rich blues of the oceans, the sweeping yellow and tan expanse of the Sahara Desert, the deep greens of rain forests in the tropics, the rust-brown sweep of the Appalachian Mountains in autumn, and the jagged, dazzling summits of the Himalaya Mountains.

No one can gaze at the Earth from over a hundred miles up and not feel awe, wonder, and amazement at the beauty of the planet and its setting in the stark blackness of space. I was glad I had studied geography intensively before launch, and when I returned I was eager to know more

Baltimore, Maryland as seen from orbiter Atlantis on shuttle mission STS-98. (NASA)

about the places I had glimpsed from orbit. That interest has become a life-long pursuit.

9. When can you see the International Space Station from Earth?

The ISS appears as a brilliant star passing across the heavens. Near sunrise or sunset, it's easily seen with the naked eye, so binoculars or a telescope are not required. The ISS is an inspiring sight—think about the six human beings living on that steady, shining point of light. To find out when and where to see it from your location, use the NASA Spot the Station website: spotthestation.nasa.gov.

10. Do the Moon and planets look closer when seen from space?

From low Earth orbit, the Moon looks no closer than it does from the ground, but its color is a stark white and gray rather than the pale yellow disk we see from the ground.

Astronauts on Earth's night side can spot five planets with the naked eye: Mercury, Venus, Mars, Jupiter, and Saturn. From the International Space Station, just 240 miles (390 kilometers) up, the planets appear much as they do from Earth under a clear night sky. Because even the nearest is 24 million miles (38.9 million kilometers) away, the planets look no closer than they do when viewed from Earth.

Once your eyes adapt to the darkness on Earth's night side—it takes about a half an hour or so—Mars will appear redder and Saturn will appear yellower in the star-filled sky than they do from the ground. You can see those same colors, plus brilliant Venus, speedy Mercury, and magnificent Jupiter, if you check them out with binoculars or a small telescope from your own backyard.

11. How is Earth's aurora created, and can astronauts see it from space?

The fantastic light shows of the aurora borealis or aurora australis were some of the most amazing sights of my space journeys. These nighttime light displays above Earth's polar regions are the result of the interaction between the solar wind—charged particles hurled out from the Sun—and our planet's thin upper atmosphere.

The solar wind, made up of charged protons and electrons, follows the Earth's magnetic field lines down toward our atmosphere near the planet's magnetic poles. When they strike and excite nitrogen and oxygen atoms in the atmosphere, about 60 miles (100 kilometers) up, these excited atoms capture electrons or collide with neighboring atoms and emit light in the process. Excited oxygen atoms emit mostly green light, and nitrogen gives off a red glow. The orbit of the International Space Station takes it close enough to Earth's auroral regions that astronauts sometimes feel as if they are skimming just above or passing through these undulating, shimmering curtains of light.

The robot arm of the International Space Station glides above a glowing auroral curtain during Expedition 32. (NASA)

12. Is it hard to take photos of the Earth from space?

From the International Space Station the Earth fills half the sky, so the planet is always offering up visual treats. Astronauts receive years of photography training from NASA experts so they can capture scientific imagery and share with us the beautiful views from orbit. With the spacecraft moving at 5 miles per second (8 kilometers per second), astronauts must use a fast shutter speed of at least 1/500 second to prevent blurring of the landscape below.

Because a sunlit scene often includes lots of contrast—from dazzling white clouds or snowfields to darker green and tan surface features—exposure control (the amount of light let into the camera) is very important. Astronauts use a camera's built-in light meter to set the lens opening correctly. A good technique is to shoot several frames of the scene, varying the exposure, to make sure at least one is properly exposed.

The author operates a 250mm Linhof camera in orbit. (NASA)

13. What kind of camera equipment do astronauts use?

Astronauts use digital rather than film cameras aboard the International Space Station for a number of reasons: cosmic radiation degrades film stored for more than a few weeks in orbit, digital cameras don't need fresh film to be launched from Earth, and digital images can be transmitted back to Earth within hours.

Space station crews use professional-grade, single-lens-reflex camera bodies along with a selection of wide-angle and telephoto lenses. Astronauts review their Earth photos on laptop computers and send the

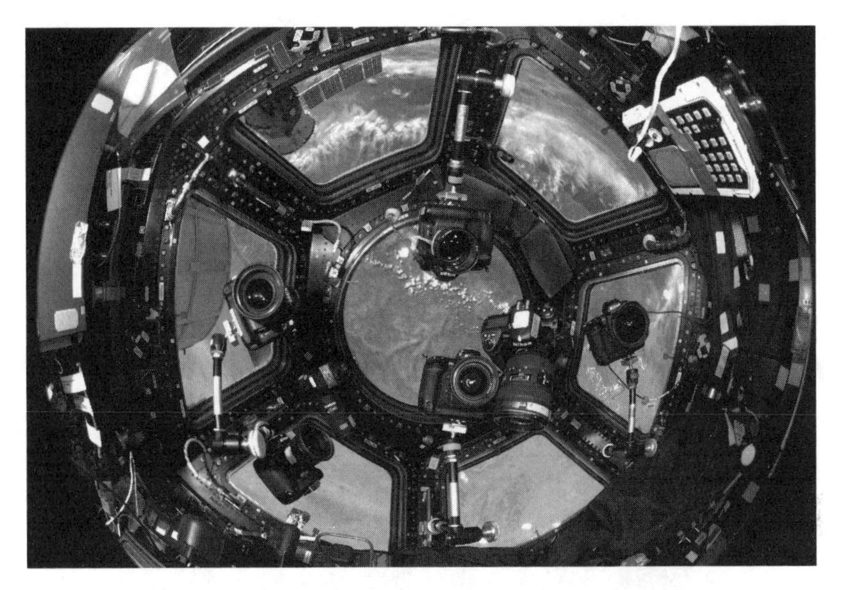

Cameras at the ready in the International Space Station's cupola, a superb Earth-viewing perch. (NASA)

best via radio link to Earth observation specialists in Houston. Cargo ships bring up replacement camera gear, and they return broken equipment and backup digital image files to Earth.

14. Can photos capture the full beauty of the Earth seen from space?

The eye will always be better at capturing the sweeping view and the wide range of constantly changing lighting conditions seen from a spacecraft window. But some cameras come close. IMAX-format films, when projected on a giant screen, present a breadth and brilliance that come close to the view enjoyed by astronauts.

Digital cameras can capture detail at night that even an astronaut's eye would miss, such as the beauty and drama of city lights and the aurora. Missing from photos of Earth are the subtleties of the planet's color palette, the delicate orange of the airglow layer shining near the top of the atmosphere, and the three-dimensional depth of atmospheric features like thunderstorms, hurricanes, and volcanic plumes. Over a million digital images of Earth have been returned by astronaut crews aboard the shuttle and International Space Station, and you can browse them at NASA's Gateway to Astronaut Photography of Earth website: eol.jsc.nasa.gov.

The beauty of this famous Apollo 17 image of the full Earth seen from near the Moon is one reason humans want to personally take in the view from space. (NASA)

The eye of Typhoon Maysak as seen from the International Space Station. (NASA)

15. Did astronauts take any close-up orbital photos of their Apollo landing sites?

The Apollo astronauts in orbit around the Moon did not have cameras powerful enough to capture details of their lunar module, equipment, and flags down at the landing sites. The robotic Lunar Reconnaissance Orbiter spacecraft has since taken images of these historic Apollo touchdown sites, showing us rover tracks, lunar module descent stages, scientific instruments, and trails of astronaut footprints. The American flags left behind can just be seen, but their nylon fabric has been bleached mostly white by the Sun's intense ultraviolet radiation.

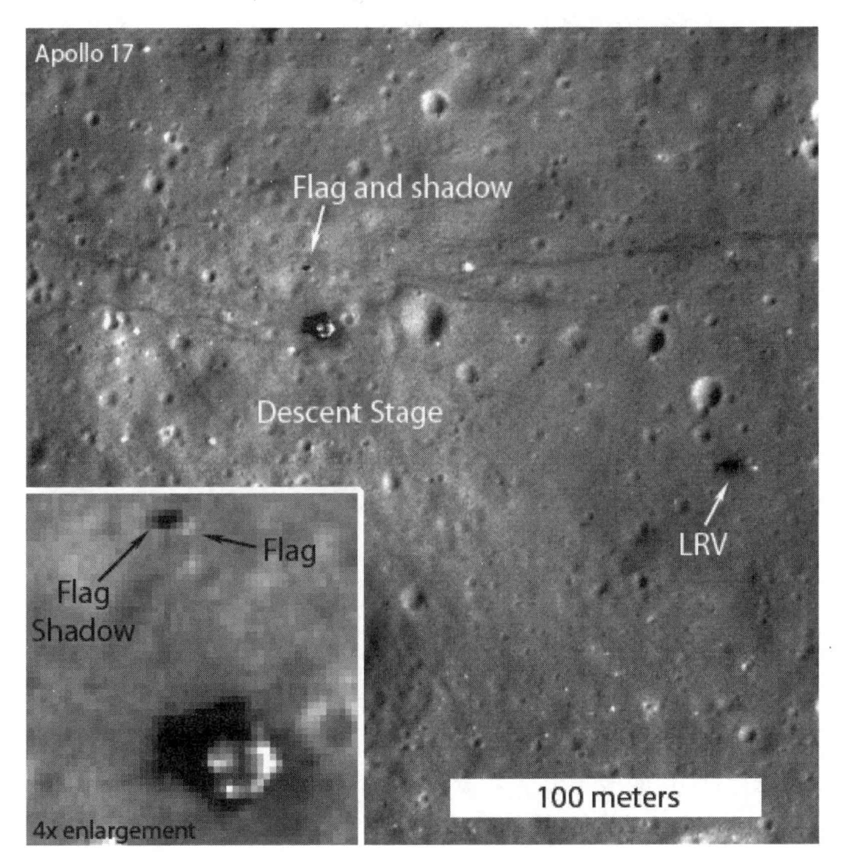

Lunar Reconnaissance Orbiter image of the Apollo 16 landing site and the American flag erected there in 1972. The dark object is the lunar module descent stage, and the LRV is the parked lunar roving vehicle. (NASA)

16. Have astronauts seen aliens in space, and are they out there?

Astronauts have not seen any evidence of alien life. Reports of unidentified flying objects (UFOs) in images returned by our spacecraft have turned out to be ice crystals, drifting orbital debris, the planet Venus, or meteors streaking through the dark atmosphere below. So far, our search for extraterrestrial life—and intelligent beings in space—has turned up no proof of alien civilizations.

Astronomers have discovered over a thousand planets around other stars, and there are at least 100 billion stars in our Milky Way. So the odds are good that life, even intelligent life, exists elsewhere in our galaxy. I think we're likely to find simple life forms in our own solar system, perhaps on Mars or on the moons of the giant planets.

As to alien civilizations, our understanding of physics tells us that it's simply too hard to travel between the stars. Even if they could, would aliens want to travel to our rather unexceptional Sun? If there are extraterrestrials out there, we'll discover them by searching for alien microbes in likely abodes in our solar system, and by listening for radio or laser signals from distant civilizations.

This artist's concept depicts Kepler-186f, the first validated Earth-sized planet to orbit a distant star in a habitable zone. (NASA Ames/SETI Institute/JPL-Caltech)

17. Did your feelings about Earth change after seeing the planet from space?

Spaceflight reinforced my view that we humans must all be responsible stewards of our home world, which nurtures our society and will support our children and future generations. I've always enjoyed studying Earth's dynamic geology, varied environments, and life's incredible diversity. I love the outdoors, too, and eagerly explore the globe's many natural wonders. Drinking in the view of Earth from space just increased my hunger to get out there and explore our home planet in detail.

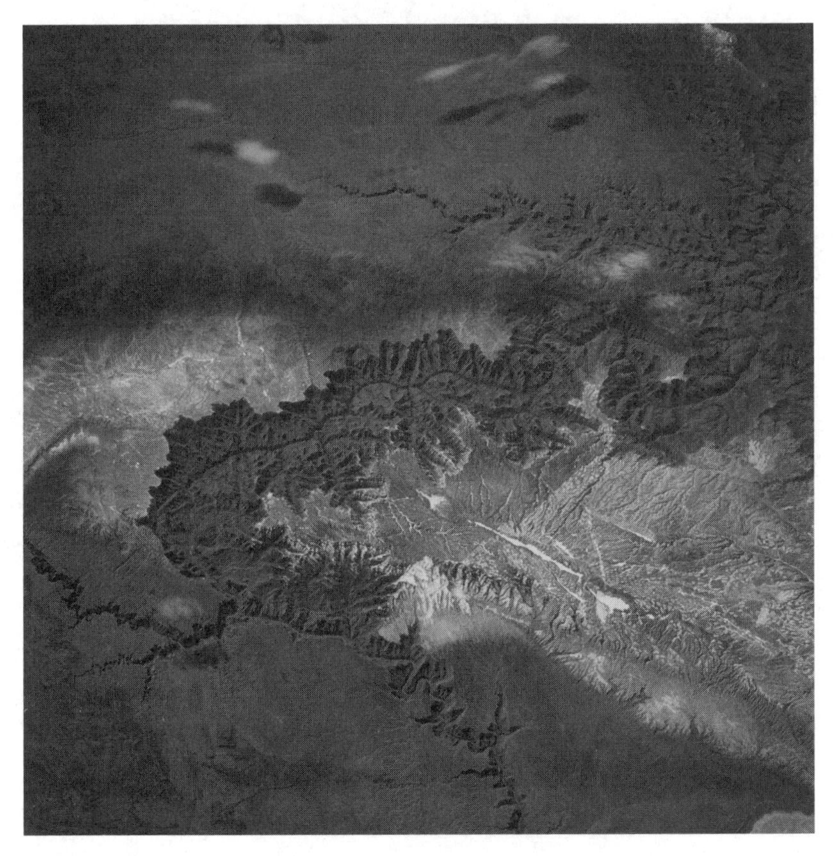

Arizona's Grand Canyon viewed from shuttle Discovery. The Colorado River splits the snowy forests of the Canyon's south rim (left) and north rim (right). (NASA)

CHAPTER XI

Returning from Space

Shuttle Atlantis prepares for its return to Earth from the International Space Station during the STS-98 mission. (NASA)

1. How do astronauts return safely to Earth?

To stay in low Earth orbit a spacecraft has to maintain a speed of about 17,500 miles per hour (28,350 kilometers per hour). When astronauts are ready to return from the International Space Station on a crew transport craft, all they have to do is slow their spacecraft by about 200 miles per hour (325 kilometers per hour). Firing retrorockets delivers the kick that slows the ship and puts it on a path back into the atmosphere, where atmospheric drag will slow the craft even further. This is called reentry.

Still traveling at about 25 times the speed of sound, the spacecraft collides with molecules in the upper atmosphere, creating a strong, highly compressed shock wave in front of the ship. Heat radiated from this shock wave, along with frictional heating from the air molecules hitting the ship's skin, raises the surrounding air temperature to more than 3,000° F (1,650° C). The ship's heat shield protects its physical structure and crew as the spacecraft's kinetic energy of motion is converted into heat energy and released into the atmosphere. Finally, the ship slows to a speed low enough to either open parachutes (like the *Soyuz*, *Crew Dragon*, or *CST-100 Starliner*) or glide to a runway landing (like the space shuttle or the proposed *Dream Chaser* space plane).

2. What preparations did you make for your reentry?

My crews devoted several hours the night before reentry to stowing all our science gear, putting away housekeeping and meal supplies, and setting up the cabin for our return. We installed the orbiter's folding seats and prepped the spacesuits for use the next morning.

On landing day, we began our landing checklist some four hours before igniting our retrorockets to start our return. Some of us taped sensors to our chests to record our pulse and the heart's electrical rhythms, along with an arm cuff to measure blood pressure during reentry. After breakfast, we took turns suiting up. Two of us helped each crewmate into his or her bulky, orange Advanced Crew Escape Suit and parachute harness. Helmets and gloves would come later.

About an hour before landing, each of us drank some 40 ounces (five cups, or a little over one liter) of warm chicken broth, or water with salt tablets, to replenish our blood volume and keep us from feeling faint during reentry and landing.

3. Is reentry as exciting as launch?

My reentry experiences were unquestionably more visually exciting than launch, for several reasons. During launch, our flight deck windows were pointed up at the empty sky. During reentry, these same windows revealed amazing views of the rapidly approaching Earth and the blanket of hot, glowing plasma that engulfed our ship as it collided with air molecules in the Earth's atmosphere. The light show through our front, side, and top windows lasted for more than twenty minutes.

Unlike launch, reentry was completely silent and almost vibration free, except for the last five minutes of buffeting as we slowed through the sound barrier nearing the runway. On the shuttle, we experienced just 1.7 g's of deceleration forces during reentry (just less than twice our normal weight), yet that head-to-toe force lasted for ten minutes, quite a strain after living weightless in free fall for a couple of weeks. My shoulder and arms sagged under that seemingly crushing weight as the feel of "almost normal" gravity returned.

Finally, the shuttle made an exhilarating spiral dive over and down to our landing site, followed by an eye-popping, nose-down plunge toward the runway. In the final seconds, our commander executed a graceful, precision pullout to a gentle touchdown—an always impressive feat of airmanship!

4. What did you see during reentry?

As the space shuttle streaked into the atmosphere at Mach 25 (25 times the speed of sound), its reentry heated the surrounding air to more than 3,000° degrees F (1,650° C), wrapping us in a glowing cocoon of incandescent plasma.

Out the front windows it blazed a neon orange, out the side hatch it shone bright pink and cherry red, and above the overhead windows flickered a white-hot trail of gas streaked with yellow and purple. Sparks flew back from the nose and joined this pulsing comet tail, lighting up the cabin in an eerie glow. Out the side windows, we took in a stupendous view of Earth racing past as we arrowed toward our landing site at Mach 20! We zoomed past airliners below us as if they were standing still.

The light show was over by the time we slowed to a speed of Mach 10. At this point we began a steep, heart-stopping rush toward our landing site. In our plunge from 15,000 to 2,000 feet (4,500 to 600 meters), we had a brief, nose-down glimpse of the runway before the commander expertly raised the orbiter's nose and settled the landing gear gently onto the concrete.

5. What did you feel during reentry?

After our retrorockets dropped us out of orbit, we fell through empty space for some 20 minutes until slicing into the top of the atmosphere. Still falling silently at Mach 25 (25 times the speed of sound), the shuttle began to heat externally as it experienced the first wisps of air.

This drag force soon had us again experiencing Earth's gravity, and for the first time in two weeks I could feel my body settling into the orbiter seat. As the shuttle slowed from Mach 25 to Mach 10 we felt no vibration, only a smooth, head-to-toe deceleration force that felt like giant hands were pressing down on our shoulders. But once in the lower atmosphere, we heard the roar of the slipstream—the air rushing past the cabin—and felt a heavy shaking as the shuttle orbiter decelerated through the speed of sound. My crewmates called this "the rocky road" back to Earth.

6. Why is the reentry angle important as the ship reaches the top of the atmosphere?

When returning from low Earth orbit, your flight path angle—the angle between horizontal and your flight path—is very important. At speeds of over 17,000 miles per hour, if your flight path angle is too shallow, you will skip back out of the atmosphere. You'll come down again, of course—Earth's gravity won't let you go—but you will miss your intended landing site. If you come in from the Moon at too shallow an angle, you'll skip off the atmosphere into a very high Earth orbit; and you could very well run out of oxygen and supplies before you encountered Earth again.

If your angle is too steep, then your spacecraft penetrates into the thicker layers of the atmosphere too rapidly, causing the craft to slow down too quickly and overheat. The spacecraft might be crushed, or burn up.

Soyuz craft leaving the ISS set up a reentry angle of -1.35 degrees. On the shuttle, we aimed for a flight path angle when hitting the atmosphere of -1 to -2 degrees. Returning from the Moon, the *Apollo* spacecraft aimed for a flight path angle of about -6.5 degrees.

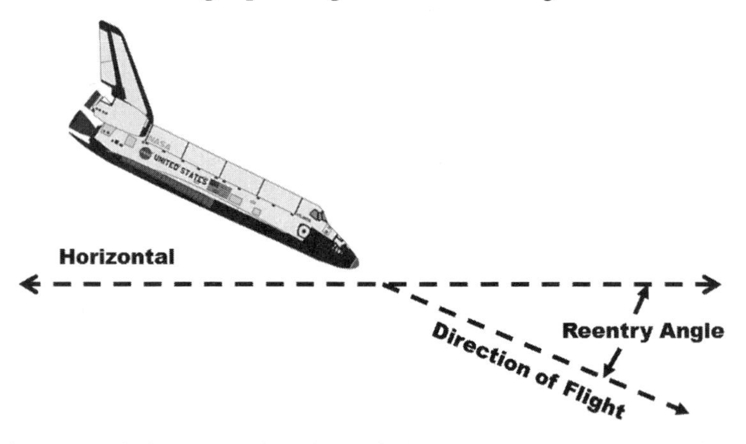

The reentry flight path angle is the angle between the horizontal (Earth's surface) and the direction the spacecraft is traveling. (Author)

7. Why does the outside of the ship get so hot during reentry?

Spacecraft coming back into the atmosphere at 25 times the speed of sound—or more, if returning from the Moon or planets—slam into air molecules in the upper atmosphere. Those molecules cannot flow smoothly around the ship—most can't get out of the way at all. These hypersonic collisions set up a highly compressed shock wave. The molecules are torn apart and their atoms are heated to very high temperatures, sometimes over 3,000° F (1,650° C). Friction from the flowing air also contributes about 20 percent of this reentry heating.

The tremendous heat strips atoms of their electrons and creates an incandescent plasma that flows around and envelops the spacecraft, radiating heat into its metal skin. Without a heat shield, the superheated plasma would melt any unprotected surfaces and destroy the ship and its crew.

Superheated reentry plasma glows outside the shuttle's flight deck windows during atmospheric reentry. (NASA)

8. How are spacecraft protected from the heat of reentry?

Piloted spacecraft today usually use ablative heat shields. These are made of a plastic-like resin that chars, melts, and vaporizes under the heat of reentry, thus carrying the heat away from the spacecraft's skin and structure. Because so much of an ablative heat shield is burned away during reentry, it can't be reused.

The space shuttle was protected by reusable tiles made of silica ceramic. They were as light as Styrofoam, and were very good at insulating the metal skin from reentry heat. The portions of the orbiter's exterior that were exposed to the highest temperatures, like the nose and leading edges of the wing, were made from reinforced carbon-carbon. The *Dream Chaser* space plane and the unmanned X-37B mini-shuttle still use this type of reusable tile. Reusable heat shield tiles, however, are somewhat fragile and require costly inspection and even replacement if damaged.

Orion's charred ablative heat shield after the spacecraft's 2014 test flight. (NASA)

9. Were shuttle landings a rough experience?

My four shuttle touchdowns were no bumpier than a good airliner landing. My commanders had practiced over a thousand shuttle approaches in the simulator and shuttle training aircraft, learning from expert NASA instructors and veteran shuttle commanders. On one of my flights, our commander landed our orbiter so gently at Kennedy Space Center that I could not tell we had touched down until the computer display in front of me showed our flight software had switched to "on the ground" mode.

Orbiter Atlantis brings my crew home from the International Space Station on space shuttle mission STS-98. (NASA)

10. What kinds of problems have spacecraft encountered on reentry?

In 1962, a faulty sensor aboard John Glenn's *Friendship 7* spacecraft indicated that his heat shield was loose. It was a false alarm, but flight controllers were very concerned that he might be incinerated during reentry. In 1967, cosmonaut Vladimir Komarov's *Soyuz 1* tumbled out of control during reentry. His parachute lines tangled when opening, and he was killed upon impact. In 2003, hot plasma penetrated damaged heat shield panels on shuttle *Columbia's* left wing, causing a structural failure and the loss of the crew. In 2008, the *Soyuz TMA-11* service module failed to separate cleanly from the crew's descent module, causing the ship's guidance system to subject the crew to a steep and punishing reentry. There was heat damage to the spacecraft, but despite smoke in the cabin and a hard landing all three astronauts landed safely. That re-entry was a little too exciting.

11. What is the typical reentry process for the *Soyuz* or other International Space Station transports?

Some three and a half hours before landing, the spacecraft un-docks from the ISS. Two and a half hours later the crew fires retrorockets to drop it out of orbit and into the atmosphere. This deorbit burn, as it is called, takes about four-and-a-half minutes to slow the craft for reen-try. About 30 minutes later, explosive bolts fire with a series of sledge-hammer blows as the unoccupied orbital and instrumentation modules separate from the descent module carrying the crew. Shortly after, the descent module enters the upper atmosphere 400,000 feet (122,000 me-ters) above the Earth's surface. After enduring the intense heat of reentry, the *Soyuz* deploys a small parachute, called a drogue, which stabilizes the craft 15 minutes before landing.

The critical next step happens at 28,000 feet (8,500 meters), when the main parachute opens to slow the craft for landing. That gives the crew quite a jolt! If the main chute fails, there is a backup, because the crew's survival depends on having one good parachute. To cushion the ground impact, the *Soyuz* fires two clusters of retrorockets a second be-fore touchdown. But despite shock-absorbing seats and crushable, cus-tom-molded seat liners, astronauts still experience a hard smack upon return to Mother Earth on the plains of Kazakhstan.

Using parachutes, the *Crew Dragon* space taxi will splash down into the ocean, and the *CST-100 Starliner* will hit Earth solidly on a set of airbags. Each will still deliver a solid jolt to crews upon landing.

12. Are wings or parachutes best for returning from space?

There are pluses and minuses for each system. A winged vehi-cle brings you back to a gentle runway touchdown, causing minimal damage to the ship and crew. However, such a space plane requires a large heat shield, and its landing gear, hydraulic flight controls, and

Six small retrorockets fire a second before landing to help reduce impact forces on the Soyuz spacecraft. (NASA)

braking mechanisms add complexity and weight. So wings and wheels are expensive.

Small, piloted craft like the *Soyuz, Crew Dragon*, and *CST-100 Starliner* use one to three parachutes to ensure a safe landing. Parachutes are simple and reliable, but winds and descent speed can still deliver a serious jolt at touchdown. One astronaut reported that *Soyuz* hit the ground "like a dumpster full of bricks." He stopped talking before impact so he wouldn't bite off his tongue. Ocean splashdowns usually provide a gentler touchdown, but they also soak the craft in salt water, corroding the craft's electronics and lightweight metal hull, which may affect reusability. NASA chose a water-landing option for its *Orion* spacecraft to eliminate the weight of soft-landing rockets or impact-cushioning airbags.

A test version of the parachute-lowered Orion spacecraft splashes down into a water tank at NASA's Langley Research Center, Virginia. (NASA)

13. How did your body react to returning to Earth gravity?

During reentry, deceleration forces caused my spacesuit to weigh heavily on my arms and shoulders, and my helmet to sag forward toward my chest. Back in full Earth gravity, after nearly three weeks in weightlessness, I felt extraordinarily heavy and sluggish. After landing, my arms and legs felt like lead, and it took nearly all my strength to rise from my seat and exit the orbiter hatch with some welcome help from the ground crew.

After about thirty minutes my weight seemed normal again, but now my problem was balance. I could only walk with feet spread wide in a careful, deliberate gait. Walking a straight line down the center of the hallway was no easy task, and I sometimes caromed off the wall as I turned a corner. Retraining my muscles took time, too. After one landing, when I was handed a glass of cold, purple grape juice in the clinic,

I forgot to squeeze the glass hard enough and dropped it right onto the new carpet.

My ability to bend and pick up dropped items without toppling over didn't return for another couple of days. Altogether, my balance and equilibrium took about three days to recover.

Two days after my *Columbia* landing, I tried to drive to the space center to review our crew's Earth photography. I made it as far as the end of my street, three houses down, when I ran over the curb trying to make my first right turn. I parked the car there, walked carefully back to the house, and asked my wife to drive me over to the photo lab.

14. How do astronauts feel after returning from a months-long expedition to the International Space Station?

As their bodies readjust to the familiar but tiring pull of Earth's gravity, returning astronauts experience a variety of symptoms, including fatigue, headache, pallor, dizziness, sweating, nausea, and sometimes even vomiting. Some of these uncomfortable but temporary conditions may last as long as a week. Understandably, astronauts spend a lot of time their first week back taking naps and reclining in a lounge chair as their body systems readapt to Earth's gravity. One astronaut told me that during a debriefing session his first week back, he fell asleep and slumped over sideways on a couch—right in front of his audience of engineers!

15. What was your favorite mission?

I think each of my four flights built on the experiences, trials, and joys of the previous ones. After two Earth observing missions on *Endeavour* (STS-59 and STS-68), I was privileged to fly a unique satellite-launching mission aboard *Columbia* (STS-80). On that flight, our crew set a record for the longest shuttle mission—18 days. These flights were followed by a mission on *Atlantis* (STS-98) that included a challenging International Space Station construction job and three spacewalks.

That last *Atlantis* mission encompassed almost everything an astronaut could experience on a shuttle flight: ascent, rendezvous, docking, robot arm work, spacewalks, visiting and building a space station, meeting the ISS crew in orbit, and an amazing return to Earth. Any single one of these missions would have been the highlight of my professional career; how could I choose just one as my favorite?

Orbiter Atlantis seen after undocking from the International Space Station on the author's STS-98 shuttle mission. (NASA)

CHAPTER XII

Back on Earth

A US Navy ship retrieves the Orion spacecraft after splashdown from its initial test flight in 2014. (U.S. Navy)

1. What did you enjoy most when you returned from orbit?

Right after landing, I enjoyed the rich smells of Earth, the whisper of a passing breeze, and the warmth of the Sun on my face. Next I looked forward to my reunion with my family. It was wonderful to resume our normal lives as a family after months of intense training and a mission full of heightened stress and danger. Other things I relished: a long, luxurious hot shower; my first night of uninterrupted sleep; a pizza, fresh fruit, and a cheeseburger.

2. How do astronauts get home after returning to Earth after a mission?

Astronauts completing their International Space Station expedition in a *Soyuz* spacecraft land on the steppes of Kazakhstan. After an initial medical evaluation, the crew is helicoptered by the Russian recovery forces to nearby Karaganda Airport. From there, Russian cosmonauts return to Star City just outside Moscow.

American astronauts board a NASA jet at Karaganda with a flight surgeon and support crew. The plane makes a direct return to Houston while the astronauts recline in comfortable lounge chairs. Flight surgeons don't want them to readapt until they undergo a thorough examination at Johnson Space Center. Less than 24 hours after returning from space, the astronauts are back at crew quarters at Johnson Space Center for a thorough round of medical tests before being reunited with their families.

When astronauts in commercial spacecraft splash down near U.S. shores or land on American soil, they will get a quick check-up and then fly back to Houston to start their rehabilitation program.

3. What is the post-flight schedule like for returning astronauts?

Astronauts returning from the International Space Station need time to readjust to Earth's gravity. The first day back is spent in crew quarters at Johnson Space Center where they undergo an initial round of medical examinations by researchers and flight surgeons to evaluate their condition.

For the next three weeks, doctors monitor the astronauts closely as

The Orion spacecraft after splashdown in the Pacific following a test flight in December 2014. (NASA)

they are put through an intensive battery of tests and recovery exercises. At least two hours a day are spent in medical tests and rehab sessions. Some are specific to each astronaut's condition, evaluating the pace of his or her recovery. Others are scientific tests aimed at studying astronaut health on long-duration space voyages. Only light duties are assigned as astronauts rest and recover their walking and balancing skills.

Crewmembers attend daily physical rehabilitation sessions, even as they start to compose their post-flight report and review the thousands of photos they shot while in orbit. While recuperating, they provide details about their expedition to flight controllers, engineers, instructors, and astronauts preparing for their own missions. Together with their crewmates, they prepare an expedition report delivered to managers of the ISS program, both in Houston and in Moscow. Anywhere from two to six months after returning from space, ISS astronauts get a well-deserved vacation, followed by a new technical assignment within the astronaut corps.

4. What medical problems do astronauts experience after spaceflight?

On landing day, and for perhaps a week afterward, astronauts who have returned from the International Space Station feel extraordinarily heavy. Working constantly against gravity quickly tires even those astronauts who followed their orbital exercise routine religiously. Most can walk right after landing, but their gait is unsteady, and the work of walking is noticeably tiring. Because blood volume is normally low in orbit, crewmembers are usually dehydrated for several days after they return. Sometimes they get dizzy even as they drink extra fluids.

Some ISS astronauts lose up to one percent of their skeletal bone mass per month, but that deficit has been shrinking as astronauts in orbit adopt more effective strength-training exercises that apply forces to the load-bearing bones. ISS astronauts lose very little muscle mass or heart-lung capacity due to their rigorous exercise schedule, but restoring coordination and balance takes time. For months, astronaut brains have ignored signals from the balance organs in their inner ears, causing confusion back on Earth about the actual direction of "down." Nausea lasting a week or more is not uncommon.

One serious concern seen in more than half of ISS crewmembers is vision degradation—usually farsightedness—that may persist after landing. Optical exams show eye abnormalities similar to those in patients with elevated levels of fluid pressure inside the skull. Researchers suspect that when body fluids shift toward the head in free fall, they put enough pressure on the optic nerve and/or the retina to cause vision changes.

5. What kind of medical exams do astronauts receive after landing?

The post-flight checkup includes tests of heart and lung function, vision, hearing, reflexes, and balance. Astronauts returning from the International Space Station are seriously balance-challenged. To test equilibrium, astronauts stand on a platform that tilts while flight surgeons note their reaction. To determine how well I was recovering from free fall conditions, investigators used ultrasound to observe my heart's blood output as I was pivoted from lying flat to an upright position. My oxygen

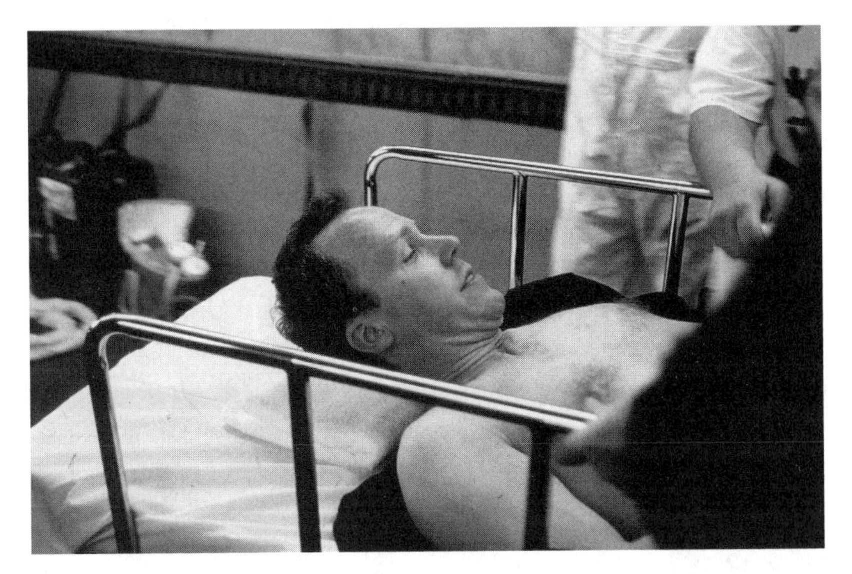

The author undergoes medical tests after landing on shuttle Endeavour, mission STS-68. (NASA)

intake was also monitored as I pedaled an exercise bike, so researchers could compare my heart and lung performance to preflight levels.

6. How long does it take astronauts to recover from a long trip to the International Space Station?

It takes between four and six weeks. After landing, astronauts embark on a program to regain prelaunch levels of coordination, stamina, and strength. Balance is a particular challenge for at least a couple of weeks. To prevent falls, they even take showers while sitting in a bathtub.

The rehabilitation program lasts for 45 days and is scheduled for two hours each crew workday. The emphasis is first on walking, flexibility, and muscle strengthening. The second phase adds balance exercises and cardio conditioning, followed by redeveloping full physical capabilities. A rehabilitation coach guides each crewmember through a well-paced program of swimming and walking in water at first, walking on a treadmill or outdoors, strength training at the gym, and finally jogging and running.

Coordination improves gradually, but it might be a month or more before an astronaut can return to favorite sports. Flight surgeons recommend astronauts not even think about their car keys for the first three weeks after landing. But the rehabilitation program is effective. Nearly everyone is released to normal duty after four to six weeks, with few if any troublesome long-term effects.

NASA astronaut Terry Virts exits the Soyuz TMA-11 after his June 2015 landing. (NASA)

7. Did you sleep normally after returning from space?

On my first night back on Earth after 18 days in orbit, I slept fitfully. In my dreams, I was sure if I lost grip on my pillow, I'd float up to the ceiling and stay there! That was one strange and restless night! But even this unsettled sleep was welcome, because by then I had been awake for close to 24 hours since I had been roused back in orbit.

8. Are there long-term negative medical effects from spaceflight?

The good news is that once astronauts get back to Earth, most body systems return to normal. Astronauts and cosmonauts so far have not suffered any apparent long-term ill effects from six months or more in orbit.

But some space-related vision changes remain after landing, and this is a concern for future crews on the International Space Station. Astronauts are tracked by NASA's health care experts for years, or even decades, after they return from space to watch for any problems that might arise. As part of this program, NASA still gives me a physical exam every year. For example, we know that being in space causes some bone loss, but we don't know if that bone loss will cause a greater risk of fractures as astronauts age. NASA astronauts who have performed spacewalks seem to have a slightly higher risk of cataracts, perhaps caused by higher radiation exposure when they are outside the spacecraft.

9. Do you get bonus pay while working in space?

One would think so, but as federal employees NASA astronauts do not get any extra pay for spaceflight. The agency doesn't award frequent flier miles either. When you rack up 22.5 million miles, as I have, that hurts! There are no automatic promotions upon return, but successful performance in space does help advance a military or civil service career. Astronauts eventually top out at the civil service GS-15 grade or the rank of colonel. Russian cosmonauts are paid less than NASA fliers, but they do receive bonuses for tasks they perform in orbit, like spacewalks, manually flown rendezvous and docking operations, or high-profile experiments.

10. How long can you serve as an astronaut?

Astronauts can remain on flight status as long as they have the desire to contribute to human spaceflight and can pass the annual flight physical exam. My crewmate on space shuttle mission STS-80, Dr. Story Musgrave, was 61 when he flew on *Columbia* with us. He is the oldest career astronaut to fly, although John Glenn, the first American to orbit the

Earth in 1962, returned to space as a shuttle payload specialist in 1998 at the age of 77. As they gain invaluable space experience, NASA often asks its veterans to move from flying assignments into management jobs. Still, many space station fliers head for orbit well into their fifties.

Blue Origin's commercial, suborbital New Shepard capsule launches on a 2015 test flight. (Blue Origin, LLC.)

11. Do all astronauts fly more than once?

If you do your job well in orbit, you are generally eligible for one or more follow-up missions. NASA wants to take advantage of the orbital experience its astronauts acquire at great expense. Flight opportunities today are not as frequent as they were during the peak shuttle years. Only about six ISS NASA crewmembers fly to orbit each year, compared to 30 or more astronauts who orbited annually on the shuttle during the 1990s. Current astronauts might have to wait 5 to 10 years for their first flight, and another 5 before they fly again.

12. What do astronauts do after leaving the astronaut corps?

Astronauts may return to military careers, resume their academic and research interests, or go to work for companies in the aerospace industry. I work as a spaceflight consultant, write books and articles, and do frequent public speaking. I am very interested in persuading policy makers of the value to be gained by exploring the Moon and asteroids: mining valuable raw materials, gaining experience in deep space operations, and preventing a future destructive asteroid impact on Earth. I also volunteer with the Association of Space Explorers, the Astronaut Scholarship Foundation, and the Astronauts Memorial Foundation.

13. Are you going back to space?

Not if I want to stay married.

14. Did you do everything you wanted to do in space?

I lived life in space at an intense pace, and tried to cram as many experiences into each day as I possibly could. If I have one regret, it's that I didn't look around enough during my spacewalks—there was too much work to do! It was hard to sneak a look at the world going by when there were so many tasks on the station still waiting to be done. Only on my final spacewalk could I afford a few minutes to drink in the awe-inspiring view of the heavens and Earth. I wish I had turned off my helmet lights during nighttime and taken in the view of the stars and Milky Way.

Overall, I felt privileged to embark on three science missions aboard the shuttle, and to fly a fourth to help construct the largest and most complex spacecraft ever built—the International Space Station. Had NASA been headed that way, I would have loved to have visited and worked on the Moon or a nearby asteroid.

15. Was it worth all the hard work to become an astronaut?

My astronaut job was the highlight of my career, and my experiences in space were the most stunning and memorable events in my professional life. I knew on my first mission that all the preparation and training it took to get to orbit was worth it. Certainly an astronaut's work is hard, but it is almost always fun. I tell people that I never worked so hard as when I was in space, but I never smiled so much, either. At the end of many long days in orbit, my cheeks ached from all the smiling I had done.

The author observes Earth from Endeavour's flight deck on shuttle mission STS-68. (NASA)

16. Are you still in touch with your astronaut crewmates?

My crewmates and I spent years together training in close quarters, and we shared an intense, amazing experience in space. Each of us put our lives in the hands of friends we came to trust completely. Because of this close connection, we instantly feel comfortable and reconnected when we meet, even years after our missions. We often meet at conferences, NASA social events, and gatherings of the Association of Space Explorers. We naturally enjoy sharing memories of our spaceflights together. These space friendships run deep, and it's hard to imagine ever losing touch with my crewmates.

CHAPTER XIII

Exploring Space and the Planets

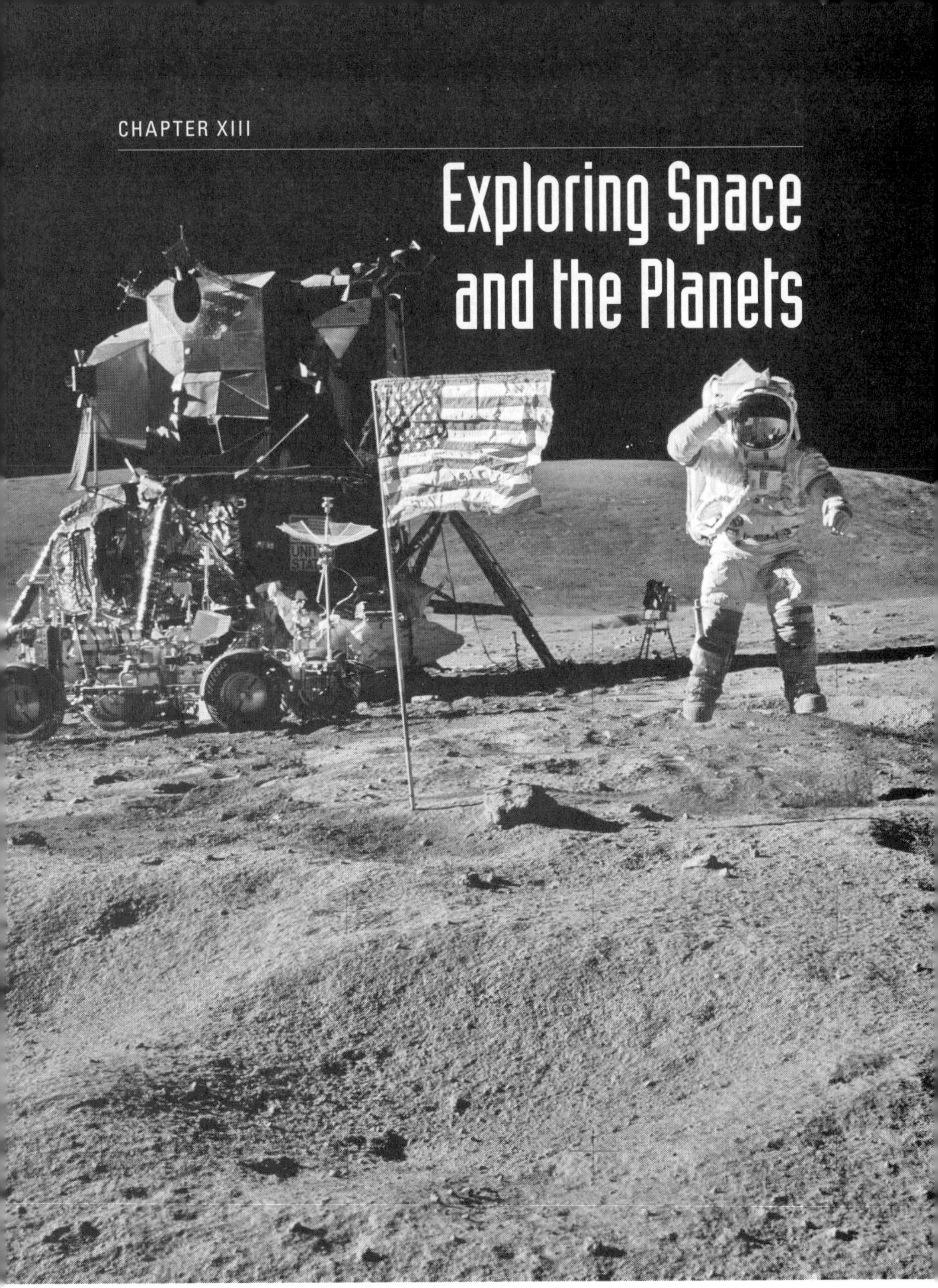

NASA astronaut John Young delivers a leaping salute to the American flag during the Apollo 16 Moon mission in 1972. (NASA)

1. Why is it important to spend money on space exploration?

Society should invest a small amount of its resources each year to create a better future. NASA's budget for space exploration is less than one percent of all U.S. government spending. This small investment improves life on Earth in a number of ways.

Exploration creates new and sophisticated technology; opens up new resources and businesses in space and on Earth; inspires young people to become the scientists and engineers we will need to tackle our future problems; expands our knowledge of our planet, solar system, and universe; and is the only way to prevent a future asteroid strike on our planet. That's a pretty good return on our investment.

2. Where should we send astronauts next in space?

The closest and most obvious destination is the Moon, which is only three days away. Exploring the Moon offers fascinating scientific opportunities as well as resources that can be used to support an outpost there. Water ice at the lunar poles and oxygen and metals from lunar dirt, or regolith, could make a lunar outpost largely self-sufficient and supply astronauts with fuel for their return trip.

Near-Earth asteroids are another attractive destination. Many come quite close to Earth, take less rocket power to reach than the Moon, and don't require a lander—astronauts can just pull up alongside these low-gravity bodies. However, a roundtrip journey to an asteroid takes a long time—four to six months at a minimum—and requires radiation shielding and reliable life support systems.

A logical path to follow would be to land robots on the Moon to explore its resources, and then send astronauts on short trips to explore the lunar surface. The same spaceships could be used to reach a near-Earth asteroid, several million miles away. In about 20 years, then, we would have enough experience to reach out for Mars.

The 1,800-foot (550-meter) near-Earth asteroid Itokawa, as seen by the Japanese Hayabusa spacecraft. (JAXA)

3. Why should we explore space when we still have problems to face here on Earth?

Exploring space will help us find new, unexpected solutions to problems that we face here on the ground. For example, a water purification system developed by NASA for the International Space Station has been adapted to provide safe drinking water to remote towns in Kurdistan. Research into keeping astronauts' bones healthy aboard the ISS has helped us understand how to treat osteoporosis, the weakened bone condition that afflicts millions of the elderly.

Space exploration can deliver the means of preventing the most fearsome natural catastrophe humanity will ever face: a devastating asteroid impact. An asteroid capable of flattening a city struck the Tunguska region of Russia in 1908, and in 2013 the explosion of a smaller one sent over a thousand people to the hospital in Chelyabinsk, Russia. We *will* be struck again by an asteroid, and space exploration is our asteroid insurance policy. We must learn to launch robot spacecraft equipped with the technologies needed to divert a dangerous asteroid by nudging it off its collision course with Earth.

In this artist's concept, a robot spacecraft slams into a dangerous near-Earth asteroid to change its orbit and make it miss its deadly appointment with Earth. An observer spacecraft monitors the resulting change of course. (European Space Agency)

4. What are some of the latest practical benefits of space exploration?

NASA's work in polishing space telescope mirrors has helped build a laser vision system that measures a patient's eye and creates a map to guide treatment. A NASA-developed scanner used to measure astronaut bone loss on the International Space Station is now used on Earth to aid early detection of osteoporosis, and space-tested techniques for reducing bone loss may help with prevention and rehabilitation of this condition.

A polymer material developed to repair the space shuttle's heat shield tiles becomes a strong ceramic when heated to high temperatures, and is now used to protect systems in the military, aerospace, and automotive markets.

A leaf sensor developed by NASA to measure moisture levels in plants grown on the ISS has been licensed by a commercial company. With this device, thirsty plants on Earth can now send text messages directly to farmers, asking for a drink.

A company that helped develop life support systems for NASA's *Orion* spacecraft now uses the same technology to build advanced suits that protect undersea divers in extreme and dangerous environments. For the latest space innovations improving life on Earth go to the NASA Spinoffs website: spinoff.nasa.gov.

5. What kind of valuable materials might be present in space?

Natural resources, such as water, oxygen, and metals found on the Moon, asteroids, and the planets are the key to making space exploration possible in the twenty-first century and supporting ourselves off planet Earth.

Water is especially valuable in space because it is composed of oxygen and hydrogen, both powerful chemical rocket propellants. Water is present at the Moon's poles, and some asteroids have water-bearing minerals on their surfaces. Oxygen can also be obtained from rocks on the Moon and asteroids.

The Mars atmosphere contains water vapor, and the *Phoenix* lander discovered water ice just a few inches beneath the surface near Mars' north polar region. The Martian polar ice caps are estimated to contain more than one hundred times as much water as there is in the Great Lakes. In 2015, the Mars Reconnaissance Orbiter detected traces of salty, liquid water trickling down the inner slopes of Martian craters. We suspect more ice is hidden in frozen soil and in glaciers buried beneath dust and rock between the planet's poles and equator.

These water sources mean that explorers can use solar or nuclear energy to manufacture oxygen and rocket fuel on the Moon, asteroids, and Mars, greatly reducing the cost of traveling to these places and surviving there.

6. What new space vehicles will be needed to explore the Moon, asteroids, and Mars?

NASA is currently testing the *Orion* multi-purpose crew vehicle, which can carry four astronauts to the Moon and beyond. Its heat shield can protect the crew from temperatures above 5,000° F (2,760° C), generated when it returns to Earth traveling at speeds faster than 25,000 miles per hour (40,500 kilometers per hour). *Orion's* next unmanned flight test will be in 2018.

To explore the Moon's surface, we'll need a lander, perhaps one that can be refueled with hydrogen and oxygen produced from lunar water. Working on an asteroid will require a one- or two-person exploration vehicle equipped with anchors and manipulator arms that can hover and glide over the low-gravity surface.

Mars is the toughest challenge. Trips there through deep space will require a powerful rocket propulsion system, radiation-shielded living quarters, and a lander with heat shield, parachutes, and soft-landing engines. Astronauts on Mars will likely refuel for takeoff with rocket propellants made from subsurface ice or water extracted from the Martian air. To return safely to Earth, the space travelers will need a reentry module like *Orion*.

7. Have you been to the Moon?

Sadly, no. I grew up with the *Apollo* astronauts as my heroes. When I was hired as an astronaut, I thought I would be a part of the nation's efforts to return to and further explore the Moon. Instead I helped build the International Space Station, where we should learn much of what we need to know to get back to the Moon and reach the nearby asteroids and Mars. Many of today's aspiring astronauts are sure to reach the Moon's dusty surface and help establish human settlements on the Red Planet.

8. Why haven't astronauts returned to the Moon since Apollo?

After the U.S. won the race to the Moon and fulfilled President Kennedy's commitment to land there and return by 1970, interest in further Apollo landings rapidly dwindled. Politicians cut NASA's budget and cancelled the final two Apollo Moon landing missions. NASA then turned its attention to flying astronauts on long-duration missions aboard the *Skylab* space station, and developing the space shuttle. In turn, the shuttle was instrumental in building and supplying the International Space Station, but that outpost is in low Earth orbit.

In 1990, President George H.W. Bush proposed returning to the Moon by 2000, but his plan was rejected by Congress. In 2004, President George W. Bush also proposed sending astronauts to the Moon no later than 2020. But neither he nor President Obama supplied the money needed to make it happen.

Now China, Russia, and the Europeans all have expressed interest in sending explorers to the Moon, and I believe the U.S. will eventually cooperate in a joint effort to follow in the Apollo astronauts' footsteps. But *when* that will happen is a political, not a technical decision.

9. Why should astronauts visit the Moon again?

The *Apollo* explorers barely scratched the surface of the Moon's rich scientific story. To learn about the early history of the Earth-Moon system, astronaut scientists could sample the ancient crust and mantle (middle rock layer) of the Moon by landing on the rim of a giant impact crater called the South Pole–Aitken basin, the oldest and deepest crater on the Moon. They could use rovers to drive hundreds of miles across the Moon, collecting rock samples, checking out young lunar volcanoes, and poking into lava caves where a sheltered base might be constructed.

To further explore the universe, astronauts could erect radio telescopes on the Moon's quiet far side. Astronauts should also exploit the ice deposits in the shadowed, super-cold craters found near the Moon's poles. These could be a source of water, oxygen, and rocket fuel. Mining lunar resources will train astronaut explorers to use the same techniques to survive on Mars.

Apollo 17's crew took this image of the Moon in 1972. The South Pole–Aitken Basin is the dark area at far left. (NASA)

10. Why would we want to mine the asteroids?

We need to learn how to extract water from the Moon and asteroids. Each molecule of water is made of one oxygen and two hydrogen atoms, so water mined in space can provide astronauts with drinking water, oxygen for breathing, and powerful rocket propellants.

A single water-rich asteroid measuring 1,670 feet (500 meters) in diameter might be worth $5 trillion, based on what it would cost to transport the same amount of water from Earth. Small asteroids have hardly any gravity, making it easy to launch water tanker spacecraft back to Earth-Moon space.

Because there are thousands of asteroids whose orbits bring them close to Earth, robots can mine several and make regular water shipments back to Earth-Moon space for fueling Mars-bound spacecraft. Asteroids can also furnish metals and organic chemicals for use in space.

Asteroids also contain valuable elements such as platinum, which might be worth bringing all the way back to Earth.

An artist's concept of the Arkyd-100 spacecraft, forerunner of a robot asteroid prospector, seen in Earth orbit. (Planetary Resources)

11. Why does NASA want to bring a nearby asteroid back toward Earth?

NASA is eager to get astronaut explorers out into deep space again for the first time since the Apollo program ended in 1972. The agency has enough money to fly the new *Orion* spacecraft and its Space Launch System rocket, but not enough to send astronauts back to the Moon's surface or to a near-Earth asteroid a few million miles away, at least not until the late 2020s. So NASA has proposed bringing a piece of an asteroid back to us.

In 2020, NASA plans to launch a solar-powered robot craft to pick up a 10- to 13-foot (3- to 4-meter), multi-ton boulder from a nearby asteroid and nudge it into a safe orbit around the Moon. Astronauts on board the *Orion* spacecraft would then rendezvous with this ancient, resource-rich asteroid fragment. On a pair of spacewalks, they would get surface samples that could be analyzed to help unravel the early history of the solar system. Their visit would lead to further experiments aimed at extracting water and valuable metals. NASA hopes this experience will help provide us with the know-how needed to reach Mars and its moons.

12. Do we need to worry about asteroids hitting the Earth?

Our Earth orbits the Sun in a cosmic shooting gallery. It constantly encounters near-Earth asteroids, rocky bodies left over from the formation of the planets 4.6 billion years ago. A hundred tons of asteroid bits and pieces hit the Earth every day. Every so often a big asteroid hits us, and it's certain a damaging asteroid will strike in the future.

But large cosmic disasters are very infrequent. An asteroid impact that would have serious global effects happens, on average, only every 100,000 years. Your odds of dying in an asteroid strike are about the same as your odds of dying in a plane crash, about 1 in 50,000.

We have the technology to find and deflect asteroids, but we have yet to invest the resources needed to avoid a disaster. Asteroid detection requires deploying better search telescopes on the ground and in space. We also need to actually nudge a harmless asteroid with a robot spacecraft to show we know how to do the job. If we can give ourselves ten years advance warning of a future impact, we should have the time to divert the asteroid from its collision course with Earth.

13. Can astronauts protect Earth from an asteroid strike?

Astronauts and robots exploring an asteroid will obtain measurements about the object's structure and physical strength to help us figure out the best way to divert a hazardous one from hitting Earth. But for the actual deflection of a rogue asteroid, we'll use robot spacecraft.

Robots can fly for several years to reach an asteroid in its orbit around the Sun, far from Earth. Their objective will be to change the asteroid's speed, and thus make it miss its fateful rendezvous with Earth. To ensure success, we'll send several robot spacecraft. Some will fly along with the asteroid as it orbits the Sun, tracking it precisely, and monitoring

A 65-foot (20-meter) asteroid disintegrating over Chelyabinsk, Russia on Feb. 13, 2015. (M. Ahmetvaleev/NASA)

In this illustration, a small gravity-tractor spacecraft flies formation with an asteroid , using its tiny gravitational attraction to "tow" the object into a safe orbit. (Dan Durda/Fellow, IAAA)

the effectiveness of our deflection efforts. Others will change the asteroid's speed by colliding with it or detonating an explosive.

The earlier and farther from Earth we carry out the deflection, the more time we have to try again if our first efforts don't work. Contrary to Hollywood movie plots, robots are the right choice for deflecting asteroids—they don't care if they're sent on a one-way trip.

14. What are the best ways to divert an asteroid?

The most promising methods of changing an asteroid's speed (and thus its orbit) to keep it from hitting Earth include:

- A gravity tractor, a robot spacecraft which hovers over a small or medium-sized asteroid and uses the tiny gravitational attraction between spacecraft and asteroid to change the asteroid's speed and orbit, making it miss the Earth.
- A kinetic impact spacecraft, which deliberately rams an asteroid at a very high speed, perhaps 6 to 12 miles per second (10 or 20 kilometers per second). The impact changes the asteroid's speed ever so slightly and causes it to miss Earth.
- A directed-energy spacecraft, which uses solar panels to power a laser beam that vaporizes a small spot on the asteroid's surface. The superheated gas generated acts like a small thruster, changing the asteroid's speed.

In the very rare case of astronomers detecting a large asteroid headed toward Earth with very little warning, a spacecraft could detonate a nuclear explosive close to the object's surface. This would create a hot plume of vaporized dust and rock that would act like a rocket exhaust to shove the asteroid in the opposite direction.

With these techniques, we can handle the most damaging asteroid collision threats, but only if we find out about them early enough to respond.

15. Why is Mars so interesting to human explorers?

Mars is the most Earth-like of all the other worlds in our solar system. Its day lasts just over 24 hours and it has regular seasons. Through a telescope we can see the effects of these seasons on the planet's surface: the shrinking and expanding of polar ice caps, and planet-wide dust storms.

Unmanned spacecraft have sent back beautiful images of its stark, desert-like landscapes. The temperatures on Mars are much colder than on Earth, but at noon on a Mars summer day they can get above freezing. With a spacesuit for oxygen and thermal protection, you'd be quite comfortable going for a short stroll outside. In the photos sent back by our robot rovers, we can imagine ourselves walking over the next sand dune or rocky ridge to see what's there.

With ice underlying much of its surface and water vapor in its atmosphere, humans will be able to make a go of living on Mars. With some hard work to develop the necessary technology, we can build a permanent outpost there, search out signs of life, and look toward a self-supporting colony.

16. How will we get to Mars?

Expeditions to Mars will require all the ingenuity and determination that humans can muster. The trips will last nearly three years, round-trip, if we use today's chemical rockets. Nuclear-powered rockets could cut the time to two years or less.

Before launching astronauts to Mars, we would first send ahead their food, supplies, a habitat module, and a return vehicle. These would be transported to Mars using slow but efficient cargo tugs. Just over two years later, astronauts would launch on a fast trip to Mars, traveling in a bus-sized module for about six months and carrying their lander with them.

Arriving at Mars orbit, the crew would land next to their already-delivered habitat. They'll have to cover it with soil to shield themselves from radiation, or robots could bury it before they got there. After about 18 months of exploration the crew would depart in their return vehicle, fueled with rocket propellant made on Mars. Their departure date would be timed to provide the quickest journey back to Earth

Once in Mars orbit again, they would dock with their cruise vehicle, transfer their Mars scientific samples, and boost out of Mars orbit for the six-month trip back to Earth. Finally, the crew's spacecraft would parachute directly into Earth's atmosphere, or rendezvous with an Earth-orbiting space station. Robots would tend the habitat left behind on Mars and prepare for the next expedition to arrive.

17. How much will it cost to send people to Mars?

Less than you might think. A recent study estimated that it would cost about $130 billion over 30 years to launch the first nine expeditions to Mars. When you take inflation into account, that is about the same amount that was spent for the ten years of the Apollo Moon landing program.

So, for the next 30 years NASA would need to spend about $4.3 billion on Mars exploration, less than a quarter of its projected budget. By comparison, in 2014 the U.S. government spent $125 billion on improper payments to citizens or on wasteful, duplicate programs. By saving just five percent of that amount each year, the nation could easily pay for the costs of human Mars exploration.

18. Will sending astronauts to asteroids help us get to Mars?

NASA has proposed astronaut visits to nearby asteroids as a strategic step toward Mars exploration. Sending astronauts beyond the Moon, to one or more asteroids, would require trips of several months or more. Such expeditions would give the agency experience with extended operations in deep space, similar to those needed for Mars expeditions.

Asteroid missions can test reliable life support machinery, the ability to operate far from Mission Control's help, and solar electric or nuclear thermal propulsion systems. NASA's first human step toward the asteroids is planned for 2025, when *Orion* astronauts will be sent to a multi-ton boulder retrieved from an asteroid and placed in orbit around the Moon.

Operating around these ancient and resource-rich bodies should teach us how to extract water and valuable structural materials, and help us reach the asteroid-like Martian moons, Phobos and Deimos. An outpost on one of these moons will give us a useful perch for eventual trips to the Mars surface.

19. How long will it take astronauts to get to Mars and back?

The total trip time would be just under three years, if you timed it when both planets were in the right orbital positions to enable a spacecraft to use the least amount of fuel. In this scenario, it would take astronauts 6 to 9 months to get there, about 18 months on the planet, and then another 6 to 9 months to return home. Shorter missions of from 1 to 2 years, which would give a crew only 30 to 90 days on Mars, would require considerably more rocket fuel. Such a trip would also force a crew to spend a year in deep space either going there or coming back.

The best way to shorten crew time in deep space on the way to or from Mars—and minimize the risks from lengthy exposure to radiation and free fall—is to use advanced rocket propulsion. A nuclear thermal rocket can cut typical cruise times in half. In this type of rocket, a nuclear reactor heats hydrogen fuel and shoots it from the exhaust nozzle at very high speeds.

20. How will we be able to feed astronauts on such long voyages?

A typical crew on a year-long, deep-space journey will need roughly 26,500 pounds of food, with no possibility of resupply along the way. If food is sent to Mars on cargo ships before the human crew arrives, it will require a shelf life of at least five years in order to retain both its nutritional value and appeal throughout the mission. Some space foods can last that long, but declining texture, color, and nutrition over time means crews would only eat them as survival rations.

Astronauts could supplement stored, pre-packaged food with crops grown in space. Some foods might be grown while cruising to Mars and others in a greenhouse on the surface. Plants being considered for space gardens include tomatoes, lettuce, spinach, carrots, strawberries, and peppers. These can be eaten raw, but other foods, like potatoes, wheat, soybeans, peanuts and rice, would require further processing before consumption.

Space gardeners will use waste from the crew to fertilize crops, and the plants in turn will absorb carbon dioxide and produce oxygen for the crew to breathe. Another important benefit from raising space crops will be the psychological lift the crew gets from working in the greenhouse.

21. Who will be the first humans to visit Mars?

We will have the technology to reach the moons of Mars in the 2030s, and to send astronauts to the surface of the planet in 2040 or shortly after. I think the explorers who take the first steps on Mars are today's young students, just getting the idea that they might one day like to work as astronauts. I was 10 when I was really bitten by the space exploration bug.

Of course, actually sending these first explorers to the Red Planet depends on us earthlings telling our representatives and leaders that exploring space is important, and deserving of healthy funding. Private industry has a vital role to play, too, using space raw materials to produce the rocket propellant, water, and radiation shielding needed for Mars expeditions.

Neither governments nor corporations can afford Mars on their own. Those nations with long experience at the International Space Station, led by the U.S., have the best chance of reaching and sustaining astronaut explorers on Mars.

22. When do you think humans will get to Mars?

Humans may set foot on the surface of Mars sometime after 2040. But we must make many advances in space technology to meet that timetable. First, we have to test the *Orion* deep-space crew vehicle and its powerful booster, the Space Launch System. Second, we must build and test a deep-space habitat with radiation shielding and life support systems. That habitat could be tested on astronaut voyages to the near-Earth asteroids by the late 2020s.

Third, we must learn to use asteroid water and dirt to create an in-space source of rocket propellant and the radiation shielding needed to protect crews on round trips to Mars. We'll then visit the Mars moons and use water we hope to find there to prepare for the first human landing on the Red Planet. Once on Mars, tapping the ice beneath the surface and the carbon dioxide in the atmosphere for water and propellants will be an absolute must if we are to explore and colonize the planet.

23. Does the Mars One group have a real plan to establish humans on Mars?

Mars One is a private group of individuals trying to raise funds and enlist volunteers to take a one-way trip to the Red Planet and establish a permanent Mars colony. The group has proposed creating a reality-TV show to follow the training and preparations for the mission, selling this as entertainment to raise money. I don't think the group is serious. Even the most successful TV shows in history would not be profitable enough to pay for a Mars expedition. Nor do we have the skills and technology yet to keep a crew safe, even on a one-way trip to Mars.

Most worrisome of all, we have neither the know-how to land astronauts there nor what they need to survive on the surface: large habitats, rovers, supply ships, excavators, and a nuclear energy plant. The Mars One proposal has at least shown the high level of interest the public has in getting humans to the Red Planet.

24. Will human explorers visit other worlds besides Mars?

Beyond our own Moon and Mars, there are few attractive solar system sites for astronauts to explore. A dwarf planet named Ceres, about the size of Texas, orbits in the asteroid belt beyond Mars. Ceres is rich in water, so it's probably worth stopping by.

Closer to the Sun, Venus has Earth-like gravity, but its surface temperature is a searing 872° F (467° C) and its atmospheric pressure is a crushing 92 times that of our world. To survive, astronauts would have to float in a balloon high in the thick, carbon dioxide atmosphere, unable to see the surface below. It would be better to send robots.

Beyond Mars, the gas giant planets, such as Jupiter and Saturn, have no solid surfaces to land on, even if we could survive their enormous gravity and powerful radiation belts. Moons like Jupiter's Europa and Saturn's Titan and Enceladus harbor liquid water oceans beneath icy crusts, but their distance from the Sun means frigid temperatures that are best handled by robots, not astronauts. The good news: Mars is equal in surface area to all of Earth's continents put together, so there is plenty of room to explore!

25. Can we live safely on other planets?

Both the Moon and Mars can supply space colonists with water, oxygen, and raw materials derived from their soil and rocks—and, in the case of Mars, its atmosphere. The Sun and nuclear reactors can provide energy to harvest these resources. The colonists would have to grow greenhouse-based food with local soil, water, and seeds brought from Earth. However, outposts on both worlds would need to be placed underground to protect colonists from strong radiation and extreme temperatures. Moonwalking or Marswalking time would be limited by the harsh radiation outside.

An open question is whether humans can remain healthy for months or years while living in a gravity field only one-sixth as strong as Earth's, as on the Moon, or about one-third as strong, as on Mars. This question could be answered by animal experiments in a centrifuge facility at the International Space Station, but that laboratory has yet to be built.

26. Do you think we will have colonies on other planets?

Today, we don't have the technology to create a self-sustaining colony on Mars, or anywhere else. Colonies will be possible once people can find a way not just to survive, but to make a living (and a better future for their children) on another world. We must learn to harvest resources in space by first mining water, construction materials, and metals from the Moon and asteroids.

The growth of commercial mining and solar energy businesses in space will provide the know-how and cash needed to finally reach—and stay on—Mars. Explorers will face many challenges as they reach for the Moon and nearby asteroids, but each problem solved will help develop the space skills needed to establish another home for humanity.

27. Do you think Pluto should be called a planet?

Pluto lost its planet status in 2006 when the International Astronomical Union formally redefined the term "planet." The new definition excluded Pluto and renamed it a dwarf planet. This was because Pluto is one of many *thousands* of similar icy objects in the Kuiper Belt, a wide region in space that stretches from the orbit of Neptune to the outer edge of our solar system. Pluto is the second largest of the many Pluto-like objects in the Kuiper Belt, and it was the first to be discovered back in 1930.

The *New Horizons* probe streaked by Pluto in 2015, and we know from the images it returned how complex and intriguing this little world is. Pluto is 1,400 miles (2,300 kilometers) across, massive enough to

Pluto as seen in July 2015 from the New Horizons spacecraft. (NASA)

have assumed a spherical, planet-like shape. It possesses five moons of its own, with Charon the largest. Due to these characteristics, and for historical and sentimental reasons, I think Pluto should have its planetary status restored.

28. When did we get our first close-up look at Pluto?

NASA's *New Horizons* probe flew past Pluto and its family of moons in July 2015, returning thousands of images of this icy, yet varied world. In these images, Pluto has a mottled appearance, with colors ranging from coal black to orange to white. The lighter areas may be water ice freshly exposed by comet impacts, with the darker areas enriched by organic materials from older collisions.

Pluto's polar caps are probably frozen nitrogen, not water ice, and most of its surface likely is covered with solid nitrogen. Frosts of methane and carbon dioxide have also been detected. *New Horizons* revealed faults, fissures, mountains, and many fewer craters on Pluto than expected. Surprisingly smooth areas on Pluto's surface show that it has been geologically active in the last billion years. There may be a liquid water ocean next to the dwarf planet's core.

29. Are there any planets in our solar system not yet visited?

No. Since the space age began in 1957, probes have been sent to different planets by several space-faring nations: the U.S., Russia, Japan, the partner nations of the European Space Agency, and India. All the major planets were visited by robot probes between 1962 and 1989, and the dwarf planet Pluto was visited by the *New Horizons* probe in 2015. Dwarf planet Ceres, the largest asteroid, was orbited and mapped by the *Dawn* spacecraft that same year.

NASA has spacecraft around the Moon, Mars, and Saturn, and one is on the way to Jupiter. In the 2020s, NASA plans to send an orbiter to probe Jupiter's moon, Europa. Now that we have made first contact with all the planets, ongoing and future planetary exploration missions are aimed at deepening our understanding of these worlds.

Dwarf planet Ceres, about the size of Texas, imaged by the Dawn spacecraft. (NASA)

30. Which planet would you go to?

Do I have to choose just one? Earth is my home and still the planet I most wish to explore. The stark, Sun-blasted landscape of the Moon fascinates me. As a planetary scientist, I'd love to explore its impact craters, its flood-plains of frozen lava, its gargantuan valleys carved by rivers of molten rock, and its mountain peaks thrown up by cataclysmic impacts.

Mars is even more interesting. It once had oceans; has plains pocked with gigantic, dormant volcanoes; and has massive ice deposits underlying its dusty, reddish surface. Its carbon dioxide atmosphere coupled with ice under its surface can be used to create and support an artificial habitat for astronaut explorers, and even fuel their rockets and rovers. The mysteries of Mars should keep us exploring there for centuries.

31. What robot exploration missions excite you?

In the last 50 years we have sent robots to every planet in our solar system. We are making ever more sophisticated robots, and we will be sending them to do exciting, difficult work: hunt for microbes on Mars, search for ice in the Moon's frigid polar craters, peek into cavernous lava tubes on the Moon and Mars; and sample the water seeping from crustal cracks on Jupiter's frozen moon, Europa.

I also want to see robot miners extracting water from the dirt and rocks on near-Earth asteroids. Water is more valuable than gold in space. We need it to make rocket propellant, to quench the thirst of astronauts on deep-space missions, and to supply businesses and factories in space.

32. Can't we do all our space exploring with robots?

Robots are getting ever-more capable, and they are cheaper to send to the planets than astronauts. Many worlds in our solar system are either so far away or have such extreme environments that it's impractical to send human explorers. I, for one, wouldn't like to spend 10 years of my life just getting to Pluto. But today's robots are neither as smart nor as mobile as a pet dog or cat, and don't have the scientific skills of a human field geologist or biochemist.

If we really want to know the answers to questions like "Is there life on Mars?" we must send humans to find out. I predict that in a couple of decades, smart robots will be partners on a human-led expedition to Mars. They will take on the tough job of working outside for prolonged periods, while astronauts apply their skills and intelligence to overseeing their progress and to other critical tasks.

33. Has NASA discovered other planets?

NASA's Kepler space telescope has greatly aided the search for planets around stars other than our Sun. These are called extrasolar

The humanoid ATLAS robot from the Florida Institute for Human and Machine Cognition steps out during the Defense Advanced Research Projects Agency 2015 Robotics Challenge. (IHMC/William Howell)

planets, or exoplanets. The space-based Kepler stared for several years at a particular patch of sky containing 150,000 stars or so. It watched each star to see if a planet revealed itself by causing the star to dim ever so slightly when the planet passed in front of its brilliant face. Kepler scientists have estimated that 22 percent of Sun-like stars have an Earth-sized world orbiting in their "habitable zone"—where the temperatures are just right for liquid water to exist on the surface.

Kepler has been credited with more than a thousand new planet discoveries. The total number of exoplanets found by ground and space telescopes like Kepler now exceeds 1,890. NASA's PlanetQuest site keeps track of the rising number of known exoplanets: planetquest.jpl.nasa.gov.

34. Have we found other planets like Earth?

Our technology for hunting exoplanets—planets around stars other than our own Sun—cannot yet see Earth-sized worlds in orbit around other stars, but we're getting close. The Kepler space telescope has discovered more than a thousand exoplanets, and eight of them are less than 2.7 times the size of Earth.

At least three of these worlds orbit their parent stars, smaller and cooler than our Sun, at a distance where temperatures allow liquid water to exist—the habitable zone. Exoplanets named Kepler 438-b and Kepler 442-b are likely to be rocky and much more Earth-like than the many hot, Jupiter-sized exoplanets first detected.

In our Milky Way galaxy of some 100 billion or more stars, there are about 20 billion Earth-like worlds. That means there's likely to be a temperate, Earth-like planet within about 12 light years of our Sun. We can see its parent star with the naked eye. Finding a true twin to Earth will have to wait for a new generation of space telescopes that can detect gases like oxygen, water vapor, and methane in exoplanet atmospheres.

35. Can we travel to the stars?

Given our current understanding of the laws of physics and visions for future technology, no human can make it to another star within a lifetime. Our most distant spacecraft, *Voyager 1*, which left the solar system in 2012, will not reach the neighborhood of another star for 40,000 years.

Scientists have proposed sending very small robots to nearby stars by accelerating them to very high speeds with beamed laser or microwave energy. They would radio back results from their exoplanet encounters. Some futurists are more optimistic: The 100 Year Starship Foundation seeks to identify the leaps in knowledge and technology needed to make interstellar flight a reality within a century.

This artist's concept compares Earth to Kepler-186f, the first validated Earth-sized planet to orbit a distant star in the habitable zone. (NASA Ames/SETI Institute/JPL-Caltech)

36. Do you think there is life elsewhere in space?

I believe we'll soon find life right here in our own solar system! Mars once had a warm, Earth-like climate, with a protective atmosphere and liquid water on the surface. Microbes from those early days might still survive in today's harsh conditions by inhabiting hot springs or in warm aquifers and rocks below the surface, where there is energy, water, and organic material—the conditions for life as we know it.

Europa, a moon of Jupiter, and Enceladus, a moon of Saturn, have oceans of liquid water below their icy crusts. In these dark oceans, warmed by the tidal tug of their parent planets, life might exist in the mix of organic chemicals deposited there by impacting asteroids and comets. If we don't find life in our own solar system, our Milky Way galaxy has an estimated eight billion Earth-sized worlds that orbit in their parent stars' habitable zones—that region where liquid water can exist on a planet's surface. I like those odds.

NASA's Curiosity rover is searching Gale Crater on Mars for evidence of environmental conditions that once could have supported life. (NASA)

37. Are there intelligent beings on other planets?

No one knows—yet. If Earth-like worlds are so common, and life is as tenacious elsewhere as it is here, there should be intelligent civilizations throughout our galaxy and the universe. Yet, in half a century of on-again-off-again listening campaigns, we have heard no signals from other celestial civilizations.

If an intelligent society was established on another world a couple of billion years ago, its signals, spaceships, and colonies should be highly visible today. So their absence in our universe, which is 13.8 billion years old, is a puzzle. It might be that becoming a technologically advanced civilization is very hard, or that alien societies like to keep quiet. Or perhaps ours is the first society to poke our noses out into space. I am optimistic that we will detect radio or laser signals from another civilization within my lifetime.

The SETI Institute's Allen Telescope Array, about 290 miles (470 kilometers) northeast of San Francisco, California, listens for intelligent signals from other civilizations. (SETI Institute)

38. Would you journey to another world in space if you knew you couldn't come back?

Thank goodness that in the time of Columbus, Ferdinand and Isabella, and Queen Elizabeth I, colonists were willing to leave their familiar lives and search for new opportunities and freedom in the New World. If a colony on the Moon or Mars offered better opportunities for my family than here on Earth, I would like to think I'd have the courage to make the leap to a new world. In the long run, some of us will need to move to other worlds and colonize them, to ensure that humanity doesn't succumb to a terrible virus or comet impact that might wipe out our civilization on Earth. To survive, we must become a multi-planet species.

CHAPTER XIV:

Big Bangs and Black Holes: Exploring the Universe

The Hubble telescope produced this view of the Gum 29 nebula, a brilliant stellar nursery 20,000 light-years from Earth in the Constellation Carina. ((STScI/AURA), A. Nota (ESA/STScI), and the Westerlund 2 Science Team)

1. How long will Earth exist?

Our Earth will survive as a planet for about another 7 billion years. We know that our star, the Sun, will slowly increase in brightness over the next 4 billion years, warming the Earth as it does.

Six hundred million years from now, as the Earth warms, its crust will absorb enough carbon dioxide from the atmosphere that plants will begin to die. About a billion years from now, the Earth's oceans will begin to evaporate so quickly that our planet will be enveloped in a warm, humid blanket of mostly water and carbon dioxide—a moist greenhouse world—and temperatures will increase even more. By 4 billion years from now, the oceans will be gone and most life will be extinct. Finally, some 7 billion years in the future, the aging Sun will turn into a swelling red giant star, enveloping and incinerating the Earth in our star's outer gaseous layers.

That's why space exploration is so important for humanity's future survival. Today, we're learning how to protect ourselves from asteroid strikes. In a distant tomorrow, we'll need to find a new home.

2. Can space exploration help make sure we don't go extinct, like the dinosaurs?

Developing our space exploration skills gives us the option of spreading humanity across the solar system, so a disaster on Earth won't wipe out our entire species. If we have outposts or colonies on the Moon, on Mars, and perhaps inside some of the bigger asteroids, the human race will likely survive a comet strike on Earth or some terrible viral epidemic that wipes out the planet's population. A multi-planet species is simply more survivable. The ultimate survival tactic, centuries in the future, would be to spread human colonists to newly discovered Earth-like planets around other stars.

3. As the Sun ages and expands, will humanity survive?

The Sun powers itself by fusing atoms of hydrogen into helium in a thermonuclear reaction, producing the light and heat for our solar system. Our star has been consuming itself in a process that has gone on for

4.6 billion years, burning about 600 million tons of hydrogen every second. At that rate, the Sun has at least 5 billion years of hydrogen fuel left.

When the hydrogen at its core is nearly exhausted, the Sun will start fusing helium, causing it to cool and swell into a red giant star. Its outer gaseous layers will swallow up Mercury, Venus, and probably Earth. In 5 billion years, humanity will have made plans to live elsewhere, likely migrating to the outer solar system.

Our Sun, now a swollen, red giant star, broils the surface of our world—five billion years from now. (Ron Miller)

4. What is a black hole and how does one form?

A black hole is a tiny but incredibly dense object in space, the remains of a massive star at least 10 – 24 times as big as the Sun. Its gravity is so intense that even light cannot escape its clutches. A black hole can form at the end of a star's life when its core runs out of fuel to burn, causing the loss of the heat and light that support its outer layers.

The dying star collapses on itself so rapidly that its materials rebound outward in a titanic explosion called a supernova. The force of this outward explosion crushes what remains of the star's core into an infinitely small point in space called a singularity. This new black hole weighs at least 2.5 times the mass of the Sun—billions of tons of matter packed so tightly together that the object now has infinite density and gravity so strong that light itself is unable to escape.

The black hole Cygnus X-1 seen in an artist's concept. It pulls hot matter from the larger star beside it. (NASA/CXC/M. Weiss)

5. If they are black, how do we know black holes are there?

Black holes are by definition invisible, but their gravity affects everything around them. Black holes were predicted by theories of stellar physics. Their existence was finally confirmed in the 1970s by instruments that could detect the radio waves and X-rays coming out of the whirlpools of super-hot gas being sucked into the hungry gravitational maws of otherwise invisible black holes.

Astronomers have also observed a pair of stars orbiting each other, but one of the stars is invisible and is three times more massive than our Sun. Only a black hole could have such mass and still be invisible. Finally, the energy streaming from the center of our Milky Way and other galaxies is so intense that it suggests there are supermassive black holes—a typical one is about one billion times more massive than the Sun—occupying the centers of many galaxies and quasars.

6. How big is a black hole?

Black holes themselves are just mathematical points, infinitely small, but their gravity influences a region several miles across from which no light can escape. The boundary of this region is called a Schwarzschild radius, or event horizon. As a black hole draws in surrounding matter and becomes more massive, its area of influence also grows. The black hole at the center of our galaxy weighs some 4.3 million times the mass of the Sun, and has an event horizon nearly as large as our solar system from which light can never escape.

7. How can we stop a black hole?

There is no way to head off a black hole—getting close to one would be very bad news. But cheer up: there aren't any in our stellar neighborhood. Black holes are not roaming the galaxy eating up suns and solar systems, and these remnants of dead stars would have to come very close to our Sun to affect Earth. Even if a distant black hole was headed our way today, it would take millions of years for it to get close enough to harm Earth. With enough warning, we could build spacecraft to enable humanity to travel to a safer solar system. The science fiction movie *Interstellar* used some wonderful special effects to illustrate what it might be like to travel near a black hole.

8. How many stars are in the Milky Way?

We live in an average-sized barred spiral galaxy, around 120,000 light years across (a light year is the distance light will travel in a year, equal to about six trillion miles). Our Sun is part of what's called the Orion Spur of the galaxy, and it sits about a third of the way out from the center. A good estimate is that the Milky Way is home to as many as 400 billion stars, the majority of which are small, dim, red dwarf stars. These are hard to detect the farther they are from Earth, so that 400 billion number depends somewhat on estimates of how many of these dwarf stars are out there. The *James Webb Space Telescope*, which can see these small, but still-warm red dwarfs shining in infrared light, should be able to give us better numbers.

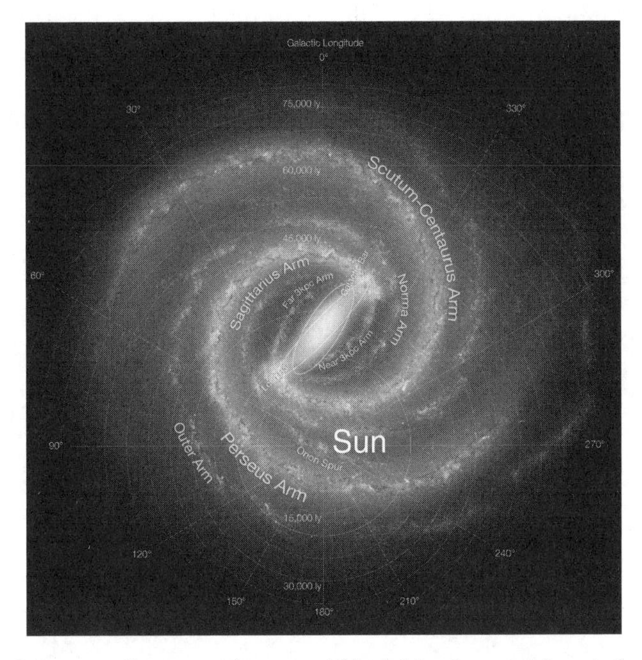

This is what our Milky Way galaxy would look like if we could see it from above. Our Sun is about a third of the way out from the galaxy's center. (NASA/Adler/U. Chicago/Wesleyan/JPL-Caltech)

9. Have astronomers discovered more galaxies besides our Milky Way?

Galaxies are clusters of hundreds of billions of stars, like our own Milky Way. Astronomers have used the *Hubble Space Telescope* along with ground-based instruments to look at very small patches of the sky. Think of *Hubble* taking in many narrow, "soda straw" views of our universe. Counting the galaxies that appear in these small patches and then multiplying that number by how many patches cover the whole sky yields a best estimate that there are at least 100 billion galaxies in our observable universe. A separate study in 2013 estimated there are at least 225 billion galaxies.

The Institute of Physics came up with a good way to visualize the size of our universe and the number of galaxies in it. If you were to hold a grain of sand up to the sky in your outstretched hand, the patch of sky it covered would contain 10,000 galaxies—each with around 100 billion stars. That's a lot of stars and planets and possible homes for life out there.

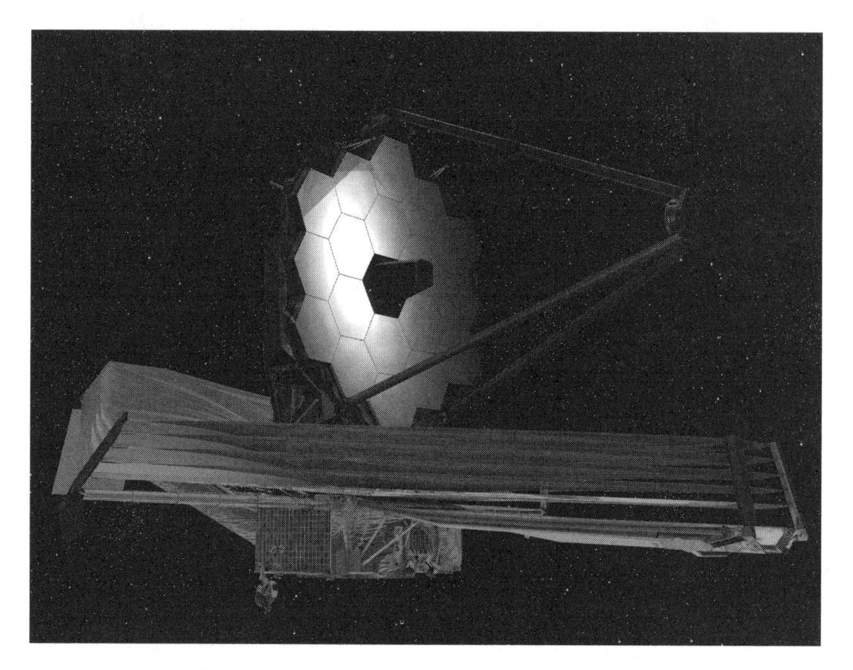

Artist's concept of the James Webb Space Telescope, the successor to the Hubble telescope. It collects infrared light with a mirror about 21 feet (6.5 meters) across. (NASA)

10. What was the Big Bang?

The Big Bang was the massive outburst of matter and light marking the origin of our present universe 13.8 billion years ago. Today we see our Milky Way and all the galaxies of the universe racing away from each other; if we trace their trajectories backward, we conclude that nearly 14 billion years ago all the matter in the cosmos was concentrated in a single point of infinite density. This incredibly dense, tremendously hot speck of energy is called a singularity. At the moment of the Big Bang, this tiny speck began to expand outward, and continues to do so today.

A second after the Big Bang and its fantastic release of energy, we think the universe was a 10-billion-degree hot sea of neutrons, protons, electrons, positrons (anti-electrons), photons (light), and neutrinos. As the universe cooled, protons and electrons combined to form hydrogen atoms, and then the process of star formation could begin. By about a billion years after the Big Bang, stars were clumping into galaxies, and giant black holes were forming at the centers of these young galaxies. As to what caused the Big Bang, no one knows. And no one knows what came before the Big Bang—our physics just can't penetrate that mystery.

11. Are there other universes?

Some physicists studying our universe believe that it is just one of an infinite number of other universes in a larger reality that's called the "multiverse." According to one theory, if space-time is infinite, we could run out to the edge of our universe and run into other versions, each slightly different from this one. Another theory is that after the Big Bang, our universe inflated, or expanded, rapidly; then, as the inflation slowed, stars and galaxies formed. But other universes may have inflated from the Big Bang and evolved differently from our own. Then there is the idea of parallel universes, which holds that other universes exist with more dimensions than the four-dimensional cosmos we live in (the three spatial dimensions, plus time).

Some scientists propose that every time we make a choice in this universe, another daughter universe is created where the opposite choice is made. In each universe, there is a version of us, thinking the choice that was made was the "real" one. Finally, mathematicians conjecture that the math in our universe is not the only type possible, and that there are alternate universes with different mathematical rules. I could go on, but I won't, because here in this universe my head is starting to hurt.

Your Future in Space

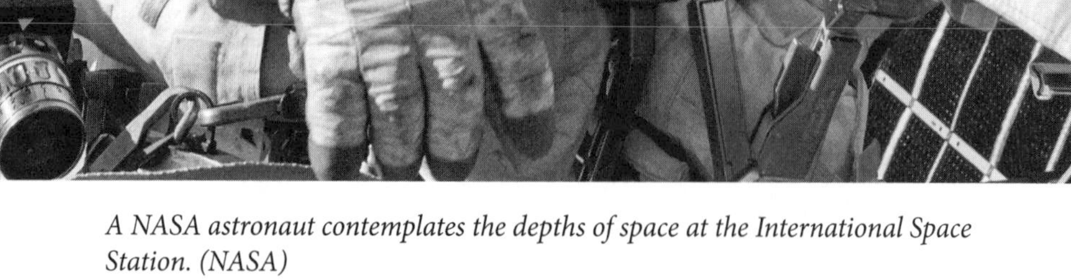

A NASA astronaut contemplates the depths of space at the International Space Station. (NASA)

1. What are my chances of getting into space?

Your chances of leaving the planet are far greater than mine were when I became interested in space travel as a child. I grew up during the Space Race, the period between 1957 and 1972 when the U.S. and the Soviet Union were competing to see who would first put a man on the Moon. The only way for an American to travel to space back then was by becoming a NASA astronaut, a process which was (and is) very competitive.

There is still a NASA astronaut corps today, whose members journey to the Space Station and one day will venture into deep space. But there are also new private paths to space. Companies like Virgin Galactic, Blue Origin, and XCOR Aerospace will offer paying customers short trips to a height of 62 miles (100 kilometers) above the Earth's surface.

Other companies, like Boeing, Sierra Nevada, and SpaceX, plan to offer space taxi service even higher into space, delivering passengers to space stations orbiting 240 miles (389 kilometers) above the Earth. SpaceX is even proposing privately-funded Mars exploration. These companies will need employees to work in space, such as space plane pilots, orbital adventure guides, hotel systems managers, and construction workers.

2. What are NASA's astronaut qualifications?

Before you can apply to be an astronaut you have to have:

- A college degree. You'll need at least a bachelor's degree in engineering, biological sciences, physical sciences, or mathematics. Good grades are very important, so study hard.

- Three years of steadily more responsible, space-related work experience. NASA seeks candidates with experience in engineering, science, or teaching; or who have at least a thousand hours as a pilot in command of jet aircraft. An advanced college degree—a master's or PhD—can take the place of some or all of the work experience.

- The ability to pass the NASA long-duration spaceflight physical. To do this you must be between 59 and 75 inches (149.5 and 190.5 centimeters) tall, have vision correctable to 20/20, and blood pressure no greater than 140/90 when you are sitting down.

The competition is fierce. Thousands apply every time NASA calls for a new astronaut candidate class. Many applicants have a master's degree or PhD in engineering or science, a medical degree, or have graduated from a professional test pilot school. Successful candidates clearly show their ability to take on and succeed at new activities and hobbies that demand judgment and teamwork.

Become an expert in your field, put your best foot forward, and then apply! More information is available at the NASA Astronaut Selection website: astronauts.nasa.gov.

3. Must I have perfect eyesight and health to become an astronaut?

To become an astronaut you have to pass the flight physical that NASA requires all its astronauts to take annually. To do so, you'll need vision corrected to 20/20, and you'll have to be in general good health with no chronic diseases or physical disabilities.

You don't have to be a perfect physical specimen. Some astronauts have been hired despite having conditions like allergies. Others have been treated successfully for cancer or have had reconstructive knee surgery. Staying physically active, participating in sports, and enjoying the outdoors help maintain good health and prepare you for astronaut training.

4. What fields of study help prepare you to become an astronaut?

You can study any scientific or engineering field, from astronomy to zoology, but it's important to choose a subject that you are enthusiastic about. Before deciding, read all about the careers you're interested in. Talk to people who work in those disciplines. Ask them what they enjoy about their work, what projects they are working on, and how their work relates to space exploration.

In college, I studied aerospace engineering, math, physics, astronautics, and space sciences. When I was in graduate school, I worked in the field of planetary science, which includes a little bit of everything that makes a solar system tick: geology, physics, atmospheric science, remote sensing, chemistry, and electromagnetism. For my research work, I used a NASA infrared telescope on Hawaii's Mauna Kea volcano to search for water on asteroids. That experience came in handy when I helped operate the shuttle-borne *Space Radar Lab* experiments, which included a radar imager to scan our planet from orbit.

5. Should I become a military or civilian pilot to become an astronaut?

No, you certainly don't need to become a military test pilot like most of the Space Race astronauts of the 1960s. The astronaut program not only seeks test pilots, but also scientists and engineers.

But NASA believes that flying builds decision-making and judgment skills that greatly improve your odds of being successful and surviving in space. That's why NASA gives its astronaut candidates without piloting experience six weeks of flight training. They then go on to qualify as crewmembers on NASA's fleet of Northrop T-38N Talons. These are high-speed, high-altitude jet trainers with advanced electronics and

NASA's Northrop T-38N Talon used for astronaut Spaceflight Readiness Training. (NASA)

navigation gear capable of aerobatics and near-supersonic speeds. Flying the T-38 regularly gives astronauts important experience in judgment and decision-making under pressure. If you choose to pursue a private pilot's license, you'll find flying to be challenging, rewarding, and fun—and you'll be ahead of the curve in preparing to work in space.

6. What are the most important steps to take if I want to work in space one day?

If you dream of going to space, it is never too early to start working toward that goal. You should:

1. Work hard and earn top grades in math and science during your years in school.
2. Choose a major in science or engineering in college, earning high marks in your chosen field.
3. When you start your career, seek work and experience in a space-related job in science or engineering.
4. Make career choices that move you closer to your goal of flying in space.
5. Be persistent and determined as you work hard in pursuit of your goals.

I've learned that no one will push you through the many steps leading toward spaceflight. You alone are responsible for getting the education and job experience needed to qualify. Many people will help you along the way: parents, teachers, professors, fellow students, co-workers, and instructor pilots. But you'll be the one to choose the goal, and you will have to do the work necessary to reach it.

Choose your own guiding star, never lose sight of it, and work hard to reach it. As Yogi Berra, the Hall of Fame baseball catcher, reputedly said: "If you don't know where you're going, there's a good chance you'll wind up someplace else."

7. How young or old can you be to qualify as an astronaut?

NASA doesn't really care how old you are as long as you have the qualifications. The youngest NASA astronaut to fly was Sally Ride, who flew at age 32. Once an astronaut, you can keep flying as long as you can pass the annual flight physical and the Chief Astronaut says you are qualified for spaceflight. For example, I flew in space with a colleague who was 61 years old.

The oldest NASA astronaut to fly was John Glenn, who in 1962 became the first American to orbit the Earth. He flew again in 1998 at age 77 on the crew of the space shuttle *Discovery*. I'm sure commercial astronauts and their space traveling passengers will one day challenge Ride's and Glenn's records.

8. How often does NASA hire astronauts?

NASA regularly hires new astronaut candidates to keep a steady stream of fresh talent coming into the astronaut corps. The agency hired a new class every two years in the 1990s when it was getting ready to build the Space Station. From 2000 through 2015, NASA hired a new class about every four years. New calls for astronauts are published on the NASA Astronaut Selection website.

Newcomers bring their knowledge of the latest developments in research, industry, and aviation into the astronaut corps, while the space veterans pass on hard-won, in-space operational experience to the candidates. The space agency wants to keep a mix of veteran and novice fliers in the corps, maintaining its space expertise as veteran fliers move on to other challenges.

9. How many astronaut candidates does NASA choose in each class?

NASA chooses as many new astronauts for each class as it needs to meet the crew size and experience demands of its upcoming missions. Over the years, class sizes have ranged from seven to 44. Recent classes have averaged fewer than 10 candidates, and it is likely to stay that way at least until 2030.

10. What jobs are astronauts assigned once they complete their training?

After their initial training, NASA's astronauts are assigned to crew positions such as commander, pilot, mission scientist, or flight engineer for service on the International Space Station (ISS) and the *Orion* spacecraft, a new spaceship designed to carry humans to the vicinity of the Moon and beyond. NASA will assign pilots and flight engineers to fly on commercial space taxi vehicles, too.

When they aren't flying, astronauts provide science and engineering support to NASA's ISS and human spaceflight program, and to fellow astronauts preparing for upcoming missions.

My astronaut class, the thirteenth chosen by NASA, had 23 members and was nicknamed "The Hairballs." (NASA)

11. What other opportunities are available for people to work in space?

We are entering the age of "personal spaceflight," when commercial space firms will at first offer quick, cannonball-style shots into space. These suborbital trips will take people 62 miles (100 kilometers) above the surface of the Earth, high enough to get a glimpse of the black daytime sky of space and the graceful curve of the Earth's horizon. Passengers on these 20-minute rocket flights will also experience about five minutes of free fall (sometimes called weightlessness). When orbital trips are offered, they will at first be very expensive; in 2015, a ticket on a Russian *Soyuz* spacecraft cost $40 – $50 million for about 10 days off the planet.

The new space tourism companies will need space taxi pilots, adventure tour guides, flight engineers, and even hotel managers. Soon after commercial tourism and space taxi companies gain a foothold, high-technology companies will start building labs and factories in space. Such firms will need skilled workers to visit, operate, and repair these facilities.

12. What if I like space but don't want to be an astronaut?

Astronauts make up a tiny fraction of the tens of thousands of people who work on the nation's space program. There are many other kinds of space jobs: engineers, scientists, lawyers, doctors, technologists, architects, technicians, flight instructors, and even veterinarians.

Some important space missions of this century won't need astronauts at all. Keeping rogue asteroids from hitting the Earth—using robots—will require a dedicated team of astronomers, mission planners, and spacecraft designers. A new generation of inventors and creative scientists will be needed to mine asteroids and the Moon, and to send skilled explorers to Mars. Space is so vast that we will never run out of frontiers to explore, and so those working to explore space will have a lifetime of chances to make new discoveries, improve life on Earth, and enable humans to live permanently on other planets.

13. Is space tourism a good idea?

Why not? Humans are naturally curious, and many of us want to personally experience the things that astronauts have described and captured in photographs and on video. Opening up space to paying passengers will put more of us in contact with a harsh yet wonderful environment. Tourism will increase our interest in seeing humans go farther "out there," to pioneer the surfaces of the Moon, the asteroids, and Mars.

As more companies compete to get passengers into space, space travel will get cheaper and safer for tourists and explorers alike. I should note, though, that space tourism companies are not exempt from the unbending laws of physics; as with all spaceflight, there is risk. Commercial space companies will have accidents, as did Virgin Galactic with its *SpaceShipTwo* in 2014. But those hard lessons will improve safety for all space travelers.

14. Can I make money in space?

Smart people are betting that there will be big business opportunities in space. In 2014, the global satellite industry had sales of over $195 billion, with $86 billion of that total earned in the U.S.

Beyond these communication and entertainment uses of space, raw materials, such as metals, water, and chemicals from the Moon and near-Earth asteroids may well become valuable commodities. Space agencies and businesses may use these natural resources to fuel future exploration and manufacturing projects. For example, a single iron-nickel asteroid 1667 feet (500 meters) across contains over 170 metric tons of platinum and related elements, worth $2 – $3 trillion dollars.

As I've mentioned, space tourism companies will try to turn a profit by taking passengers to and from orbit. Other firms will build private space stations for use in recreation, research, or manufacturing. The tax revenue generated from space industries should help pay for future expeditions to Mars.

15. How far will space exploration advance in the next two decades?

As rocket scientist Robert H. Goddard said, "It is difficult to say what is impossible, for the dream of yesterday is the hope of today and the reality of tomorrow." In the next two decades, I think explorers and pioneers will return to the Moon and travel deep into space to reach the intriguing near-Earth asteroids. Such deep-space journeys will last six months to a year.

Using raw materials already present in space is the key to helping us explore and thrive out there. Commercial firms will operate robot refineries on the Moon to produce rocket fuel, drinking water, and oxygen to supply human explorers and private visitors. Private companies will send water from nearby asteroids back to Earth orbit to be turned into fuel for satellites and spacecraft. Asteroid metals can be turned into beams and wires to build space structures and habitats, and some, like platinum, may be valuable enough to ship back to Earth.

Beyond the Moon, NASA and its partners will send pioneers deeper into space to build a supply depot on the Martian moons, Phobos and Deimos, preparing us for our first trips to the surface of the Red Planet.

But none of this will happen without the enthusiasm and ingenuity brought to space by a new generation of explorers. I hope some of my readers will be among them!

In this artist's concept, an astronaut takes a sample from the surface of a near-Earth asteroid, searching for valuable water, metals, and useful chemical compounds. (NASA)

16. May I send you another question if you didn't include it in *Ask the Astronaut*?

Of course. And thank you for your curiosity about our universe and our exciting efforts to explore its wide open spaces. I've been asked thousands of questions over the years. Unfortunately, the space in this book doesn't permit me to answer every question imaginable about astronauts and space exploration, but I'm willing to tackle a few more. Check out my website, www.AstronautTomJones.com, where I've got a page listing additional questions and answers. I'll also be answering new questions via Facebook, at "Ask the Astronaut by Astronaut Tom Jones."

Exploration is a story without an end, and an explorer should never stop asking questions. We humans are curious creatures by nature, never content to settle for what we already know. Indeed, that drive to find out what's beyond our Earth, beyond the next planet, and beyond our solar system and galaxy is one of the most important traits of *being* human. And I'm counting on a new generation of space explorers to carry forward this quest for knowledge. Will you be one of them?

The author prepares for reentry on shuttle Columbia, mission STS-80. (NASA)

Glossary

Apollo. The third U.S. human spaceflight program, which carried 12 astronauts to the surface of the Moon. Apollo spacecraft flew crewed missions between 1968 and 1975.

Association of Space Explorers. The global society of astronauts and cosmonauts who have flown orbital missions in space.

asteroid. A small rocky body orbiting the Sun.

astronaut. A person trained professionally to serve as a crew member of a spacecraft.

centrifuge. A horizontally rotating machine that subjects occupants at the end of its rotor arm to increased acceleration forces in preparation for spaceflight. It can also create an acceleration that is a substitute for gravity.

Coordinated Universal Time (UTC). The 24-hour time standard that is the basis for how time is measured and shown on clocks around the world today.

cosmic ray. A heavy, high-speed atomic particle—usually a proton or atomic nucleus—created in supernova explosions of dying stars or in active galactic nuclei found at the center of galaxies.

Crew Dragon. A piloted spacecraft built by SpaceX designed to transport cargo and people to destinations in low-Earth orbit. It is based on the unpiloted SpaceX cargo vehicle used for delivering NASA cargo to the ISS.

CST-100 Starliner. A piloted spacecraft built by Boeing and designed to carry astronauts and cargo to destinations in low-Earth orbit.

Deimos. The smaller and more distant of the two moons of Mars.

docking. The mechanical contact between two spacecraft, enabling operations such as crew or supply transfer and the assembly of space vehicles.

EMU. An abbreviation for *Extravehicular Mobility Unit*, NASA's operational spacesuit at the ISS. It enables astronauts to work efficiently outside a spacecraft.

ESA. An abbreviation for *European Space Agency.*

EVA. An abbreviation for *extravehicular activity,* a spacewalk performed outside the pressurized interior of a spacecraft.

free fall. The motion of an object under the influence of only one force, gravity. Less accurate but similar terms include weightlessness, zero-g, and microgravity.

g. An abbreviation for the acceleration at Earth's surface due to gravity.

galaxy. A large system of stars held together by mutual gravitation and isolated from similar systems by vast regions of space.

Gemini. The second U.S. human spaceflight program, which flew 10 pairs of astronauts into orbit in 1965–1966 to prepare for Apollo flights to the Moon.

ground. The term used by astronauts for the collective talents and wisdom of flight controllers and their colleagues at the Mission Control Center.

Hairballs. The self-assigned nickname for Astronaut Group 13, selected by NASA in 1990.

hypersonic. Highly supersonic speeds, usually understood as greater than Mach 5, five times the speed of sound.

International Space Station. The orbiting outpost launched and assembled by a coalition of space-faring nations beginning in 1998.

ISS. An abbreviation for *International Space Station.*

JAXA. An abbreviation for the *Japan Aerospace Exploration Agency.*

JPL. An abbreviation for the *Jet Propulsion Laboratory*, operated for NASA by CalTech in Pasadena, California. The lab specializes in robotic planetary exploration and Earth observation missions.

JSC. An abbreviation for *Johnson Space Center*, NASA's field center in Houston, Texas, specializing in human spaceflight.

KSC. An abbreviation for *Kennedy Space Center*, NASA's field center at Cape Canaveral, Florida, specializing in launch operations.

launch escape system. A safety system designed to pull astronauts clear of a failing rocket or other launch emergency.

LCVG. An abbreviation for *liquid cooling and ventilation garment*, worn by astronauts inside a spacesuit, that regulates body temperature using tubes of circulating water and oxygen.

Lynx. A commercial space plane built by XCOR and designed for suborbital tourist flights to the edge of space.

Mach number. The ratio of the speed of an aerospace vehicle relative to the local speed of sound. A vehicle traveling at the local speed of sound, for example, flies at Mach 1.

Magellanic Clouds. A pair of irregular dwarf galaxies, visible from the southern hemisphere, which may be orbiting our Milky Way galaxy.

Mercury. The U.S.'s first human spaceflight program, which flew single-seat capsules on six piloted missions between 1961 and 1963.

Milky Way. Our home galaxy, to which our Sun and solar system belong.

Mir. The Soviet, and later Russian, space station which orbited Earth between 1989 and 2001.

Mission Control. A ground control center and its personnel, responsible for the effective execution of space flights.

mission elapsed time. A method of time-keeping in space, where the mission elapsed time is set at zero at launch and counts forward in normal days, hours, minutes, and seconds. This system was prominently used during the Apollo and shuttle programs.

mission specialist. A NASA astronaut aboard the space shuttle who was primarily responsible for science and spacecraft operations in orbit, including spacewalks and robot arm operations.

MMOD. An abbreviation for *micrometeoroid and orbital debris*—natural and man-made particles in space that pose a threat to spacecraft.

Moon. Earth's natural satellite. Lower-case "moon" refers to any natural satellite of any planet or asteroid.

NASA. An abbreviation for the *National Aeronautics and Space Administration*, the U.S. government space agency.

NBL. An abbreviation for the *Neutral Buoyancy Laboratory*, NASA's mammoth, six-million-gallon water tank in Houston used to simulate free-fall behavior of spacesuits, tools, equipment, and astronauts.

near-Earth object. A small asteroid or comet whose orbit has a closest approach to the Sun equal to or less than 1.3 astronomical units, or 121 million miles (196 million km).

New Shepard. The commercial spacecraft designed by Blue Origin, intended for use in suborbital space tourism.

Orion. NASA's deep-space craft, designed to carry astronauts to the vicinity of the Moon and nearby asteroids. It may be a component of a larger Mars-bound spacecraft.

pad abort. An emergency halt of the launch countdown after the rocket's engines have ignited on the launch pad.

payload specialist. A NASA space shuttle crew member chosen to fly one or two missions to operate a particular science payload or perform a specific mission duty.

Phobos. The largest and closest of the two moons of Mars.

plasma. One of the four fundamental states of matter, the others being solid, liquid, and gas. Plasma consists of high-temperature, charged particles, and is the most abundant form of ordinary matter in the universe.

PLSS. An abbreviation for *Primary Life Support System*, the backpack-worn life support system used with NASA's current ISS spacesuit, the EMU.

Progress. An unpiloted Russian cargo ship used to resupply the International Space Station.

reaction control system (RCS). A control system on a spacecraft that uses small rocket thrusters to change the attitude (orientation) or orbit of the spacecraft, often abbreviated as RCS.

rendezvous. A spaceflight operation wherein one spacecraft approaches to within a few meters of another and stabilizes there for docking, observation, or retrieval operations.

retrorocket. A rocket which applies a force opposite to a spacecraft's direction of flight, typically used to slow a spacecraft for reentry.

remote manipulator system (RMS). The remotely operated mechanical arm—sometimes called the robot arm—on the shuttle and the International Space Station, used by astronauts and flight controllers to grapple spacecraft, move astronauts and equipment, and perform maintenance tasks.

rocket. An engine or vehicle that uses onboard propellant to create and expel a high-speed exhaust. This action creates an opposite reaction which moves the vehicle in the desired direction.

SAFER. An abbreviation for *Simplified Aid For EVA Rescue*, a backpack-worn rocket pack that enables an astronaut to return to a spacecraft after inadvertently coming adrift in orbital flight.

Salyut. The first space station design orbited by the Soviet Union in the 1970s and 1980s.

Shenzhou. The piloted spacecraft operated by the People's Republic of China. It is based on the Russian Soyuz design.

Skylab. The first space station orbited by the United States, visited by three astronaut crews in 1973–1974.

SLS. An abbreviation for *Space Launch System.*

Soyuz. The piloted spacecraft used by Russia to transport astronauts to and from the International Space Station.

space adaptation syndrome. A medical condition that causes some astronauts to experience nausea as they adapt to the free-fall environment of orbital flight. Its suspected cause is the reaction to free fall of the body's system that helps control balance.

spaceflight participant. A space traveler who has purchased a journey into space from a government or commercial space firm.

Space Launch System. A heavy-lift rocket built by NASA and designed to carry Orion and other spacecraft components into deep space.

SpaceShipTwo. A space plane designed by Virgin Galactic to take passengers on suborbital spaceflights.

space shuttle. A U.S. spacecraft consisting of a reusable orbiter space plane, reusable rocket boosters, and an expendable, external fuel tank. The shuttle operated from 1981 to 2011.

STS. An abbreviation for *Space Transportation System*, NASA's official name for the space shuttle.

sublimator. A spacesuit component that chills the water circulating through an astronaut's cooling garment by transforming solid water ice to a vapor.

supernova. The collapse and subsequent explosion of a dying star, which briefly outshines an entire galaxy.

T-38. The Northrop T-38 Talon is a U.S. Air Force two-seat jet trainer used by NASA for astronaut Spaceflight Readiness Training.

taikonaut. The term used by English-language media for a professional space traveler from China—a Chinese astronaut. It is an English-origin fusion of the Chinese *taikong* (space) and the Greek *naut* (sailor). The translation of the Chinese term for their space travelers is "space navigating personnel."

thermostabilization. The space food preservation process that uses heat and pressure to destroy microorganisms and alter the catalytic activity of enzymes in food, with the goal of preventing spoilage.

Vomit Comet. The nickname given to a series of NASA aircraft used to create free-fall conditions on Earth for the purpose of training astronauts and conducting scientific research. NASA operated a series of converted Boeing KC-135 tanker aircraft from 1967 to 2004, all sharing this nickname because the rollercoaster-like rides in these planes could make you feel sick to your stomach.

weightlessness. See "free fall."

Weightless Wonder. Nickname given to NASA's reduced gravity aircraft, a McDonnell-Douglas C-9 transport, which succeeded the KC-135 Vomit Comet in 2005. NASA retired the C-9 in 2015. These aircraft were used in astronaut training programs to simulate free fall.

Zarya. The first component of the International Space Station launched into space, boosted to orbit by Russia in 1998. It translates to English as "Dawn."

zero-g. See "free fall."

Bibliography

Apt, Jay, Michael Helfert, and Justin Wilkinson. *Orbit: NASA Astronauts Photograph the Earth.* Washington, DC: National Geographic Society, 1996.

Association of Space Explorers. *The Home Planet.* Kevin W. Kelley, ed. Reading, Mass.: Addison-Wesley, 1988.

Chaikin, Andrew. *A Man on the Moon: The Voyages of the Apollo Astronauts.* New York: Viking Press, 1994.

Chaikin, Andrew. *A Passion for Mars: Intrepid Explorers of the Red Planet.* New York: Harry N. Abrams, 2008.

Dick, Steven J., Robert Jacobs, Constance Moore, Anthony Springer, and Bertram Ulrich, eds. *America in Space: NASA's First Fifty Years.* New York: Harry N. Abrams, 2007.

ESA. The website for the European Space Agency; http://www.esa.int/ESA. Accessed July 30, 2015.

Hadfield, Chris. *An Astronaut's Guide to Life on Earth: What Going to Space Taught Me About Ingenuity, Determination, and Being Prepared for Anything.* New York: Little, Brown and Company, 2013.

Hadfield, Chris. *You Are Here: Around the World in 92 Minutes—Photographs from the International Space Station.* New York: Little, Brown and Company, 2014.

JAXA. The website for the Japan Aerospace Exploration Agency; http://global.jaxa.jp/. Accessed July 30, 2015.

Jenkins, Dennis R. *Space Shuttle: The History of the National Space Transportation System—The First 100 Missions.* Hong Kong: World Print Ltd., 2002.

Jones, Thomas D., and Michael Benson. *The Complete Idiot's Guide to NASA.* Indianapolis: Alpha Books, 2002.

Jones, Thomas D., and June A. English. *Mission: Earth—Voyage to the Home Planet.* Scholastic Press, 1996.

Jones, Tom, and Ellen Stofan. *Planetology: Unlocking the Secrets of the Solar System.* Washington, DC: National Geographic Society, 2008.

Jones, Tom. *Skywalking: An Astronaut's Memoir.* New York: Smithsonian Books—Collins, 2006.

Murray, Charles, and Catherine Bly Cox. *Apollo: The Race to the Moon.* New York: Simon and Schuster, 1989.

NASA. *Apollo Expeditions to the Moon. NASA SP-350.* Edited by Edgar M. Cortright. Washington, DC : Government Printing Office, 1975.

NASA. "A Day in the Life Aboard the International Space Station." https://www.nasa.gov/audience/foreducators/stem-on-station/dayinthelife. Accessed July 30, 2015.

NASA. "Astronaut Selection Program," accessed July 30, 2015, http://astronauts.nasa.gov/.

NASA. "NASA Science: Astrophysics," accessed July 30, 2015, http://science.nasa.gov/astrophysics/.

NASA. "NASA Science: Earth," accessed July 30, 2015, http://science.nasa.gov/earth-science/.

NASA. "Hubble Space Telescope," accessed July 30, 2015, http://www.nasa.gov/mission_pages/hubble/main/index.html.

NASA. "International Space Station," accessed July 30, 2015, https://www.nasa.gov/mission_pages/station/main/index.html.

NASA. "James Webb Space Telescope," accessed July 30, 2015, http://www.jwst.nasa.gov/.

NASA; Jet Propulsion Laboratory; "PlanetQuest," accessed July 30, 2015, http://planetquest.jpl.nasa.gov/.

NASA. "Journey to Mars," accessed July 30, 2015, http://www.nasa.gov/topics/journeytomars/index.html.

NASA. "Near Earth Object Program," accessed July 30, 2015, http://neo.jpl.nasa.gov/.

NASA. "Solar System Exploration," accessed July 30, 2015, http://solarsystem.nasa.gov/index.cfm.

Neufeld, Michael J. *Milestones of Space: Eleven Iconic Objects from the Smithsonian National Air and Space Museum.* Minneapolis: Zenith Press, 2014.

Reichhardt, Tony, ed. *Space Shuttle: The First 20 Years.* London: DK Publishing, Inc., 2002.

Roach, Mary. *Packing for Mars: The Curious Science of Life in the Void.* New York: W.W. Norton & Company, 2010.

Seedhouse, Erik. *"SpaceX: Making Commercial Spaceflight a Reality.* Chichester: Springer Praxis Books, 2013.

Squyres, Steve. *Roving Mars: Spirit, Opportunity, and the Exploration of the Red Planet.* New York: Hyperion Books, 2005.

Williams, Jeffrey N. *The Work of His Hands: A View of God's Creation from Space.* New York: St. Louis, 2010.
Yeomans, Donald H. *Near-Earth Objects: Finding Them Before They Find Us.* Princeton: Princeton University Press, 2013.
Young, Amanda. *Spacesuits: The Smithsonian National Air and Space Museum Collection.* New York: powerHouse Books, 2009.

Acknowledgments

I wish to thank my NASA astronaut colleagues who generously answered my questions about many facets of space travel. Particularly helpful were Ken Cockrell, Sandy Magnus, Tom Marshburn, Pam Melroy, Don Pettit, and Carl Walz. The editorial team at Smithsonian Books, including Carolyn Gleason, Christina Wiginton, Jaime Schwender, Matt Litts, Leah Enser, Raychel Rapazza, and Jean Crawford, offered many positive suggestions to improve the manuscript, and produced a book that would be a worthy addition to the library on the International Space Station. Service Station created an eye-catching cover design, and my crewmate Marsha Ivins took that priceless photo of me working outside the ISS. NASA's Gwen Pittman helped me locate additional photos. My agent, Deborah Grosvenor of the Grosvenor Literary Agency, helped guide this book from proposal to publication. I strove to achieve NASA-like levels of accuracy in writing this book, and any errors that slipped into the text are my responsibility.

I also thank my wife, Liz Jones, who spent many hours patiently proofreading the manuscript, and offered many, many suggestions for improving and sharpening my writing. She experienced the joyful and anxious moments of my space journeys, and never buckled. Without her, my successes in space, and with this book, would not have been possible.

About the Author

Thomas D. Jones, PhD, is a scientist, author, pilot, and veteran NASA astronaut. In more than eleven years with NASA, he flew on four space shuttle missions to Earth orbit. On his last flight, Dr. Jones led three spacewalks to install the centerpiece of the International Space Station, the American *Destiny* laboratory. He has spent fifty-three days working and living in space.

After graduation from the Air Force Academy, Tom piloted B-52D strategic bombers, earned a doctorate in planetary sciences from the University of Arizona, searched for water on asteroids, engineered intelligence-gathering systems for the CIA, and helped NASA develop advanced mission concepts to explore the solar system.

Tom is the author of *Ask the Astronaut* and other space and aviation books: *Planetology*, (written with Ellen Stofan), *Hell Hawks! The Untold Story of the American Fliers Who Savaged Hitler's Wehrmacht* (with Robert F. Dorr), and *Sky Walking: An Astronaut's Memoir.* The Wall Street Journal named *Sky Walking* one if its "Five Best" books on space. He writes frequently for *Air & Space Smithsonian*, *Aerospace America*, *Popular Mechanics*, and *AOPA Pilot* magazines.

Dr. Jones' awards include the NASA Distinguished Service Medal, four NASA Space Flight Medals, the NASA Exceptional Service award, the NASA Outstanding Leadership Medal, the NASA Exceptional Public Service award, Phi Beta Kappa, the Air Force Commendation Medal, and Eagle Scout. The Main Belt asteroid 1082 TomJones is named in his honor.

Tom served on the NASA Advisory Council, and is a board member of the Association of Space Explorers and the Astronauts Memorial Foundation. As a senior research scientist at the Florida Institute for Human and Machine Cognition, he focuses on the future direction of human space exploration, uses of asteroid and space resources, and planetary defense. He speaks frequently to audiences worldwide, and appears on TV and radio with expert commentary on science and space flight.

Information about Tom's availability as a speaker and purchasing signed copies of this book is available at www.AstronautTomJones.com.

Index

Page numbers in italics refer to photographs and illustrations.